Molecular Symmetry
and
Group Theory

Molecular Symmetry and Group Theory

A Programmed Introduction to Chemical Applications

SECOND EDITION

ALAN VINCENT

School of Chemical and Pharmaceutical Sciences
Kingston University, UK

JOHN WILEY & SONS, LTD

Chichester · Weinheim · New York · Brisbane · Singapore · Toronto

First edition © 1977 by John Wiley & Sons, Ltd
Reprinted 1978, 1979, 1981, 1983, 1985, 1987, 1988, 1990, 1992, 1993, 1996 (twice), 1997, 1998

Second Edition copyright © 2001 by John Wiley & Sons Ltd,
Baffins Lane, Chichester,
West Sussex PO19 1UD, England

National   01243 779777
International (+44) 1243 779777
e-mail (for orders and customer service enquiries): cs-books@wiley.co.uk
Visit our Home Page on http://www.wiley.co.uk

Other Wiley Editorial Offices

New York, Weinheim, Brisbane, Singapore, Toronto

Library of Congress Cataloging in Publication Data
Vincent, Alan
  Molecular symmetry and group theory

  Bibliography: p.
  Includes index.
  1. Molecular theory—Programmed instruction.
  2. Symmetry (Physics)—Programmed instruction.
  3. Groups, Theory of—Programmed instruction. I. Title.
QD461.V52 2000-10-16
541.2′2′077–dc21

                                    00–043363

British Library Cataloging in Publication Data

A catalogue record for this book is available from the British Library

ISBN   0 471 48938 7 (HB) 0 471 48939 5 (PB)

Typeset in 10/12 pt Times by C.K.M. Typesetting, Salisbury, Wiltshire.
Printed and Bound in Great Britain by Biddles Ltd, Guildford & King's Lynn
This book is printed on acid-free paper responsibly manufactured from sustainable
forestry, in which at least two trees are planted for each one used for paper production.

Contents

Preface to the Second Edition

The first edition of this book was well received by both students and teachers. The second edition, therefore, has required only minor changes to the first seven chapters. In these I have put more emphasis on the idea of the basis of a reducible representation and have clarified a few small ambiguities which reviewers have pointed out. The diagrams have also been completely re-drawn. The major addition in this edition is a completely new chapter on linear combinations. This not only introduces the projection operator method as the rigorous approach to finding the form of vibrations, wave functions, etc., but goes on to develop a simplified approach to the subject making direct use of the character table. Again the emphasis is on the application of the techniques to real chemical problems rather than on the mathematics of the method. I hope that this will give readers an enthusiasm for symmetry methods and encourage them to learn more via the excellent advanced texts cited in the bibliography.

Finally I would like to thank the (often anonymous) reviewers whose comments have been helpful in the process of revision and all the staff at John Wiley & Sons for their patience as I failed to meet various deadlines.

Alan Vincent
Kingston University
2000

How to use the Programmes

Each programme starts with a list of learning objectives, and a summary of the knowledge you will need before starting. You should study these sections carefully and make good any deficiencies in your previous knowledge. You may find it helpful at this stage to look at the revision notes at the end of the programme which give a summary of the material covered. The test, also at the end, will show you the sort of problems you should be able to tackle after working through the main text (but don't at this stage look at the answers!).

The body of each programme consists of information presented in small numbered sections termed *frames*. Each frame ends with a problem or question and then a line. You should cover the page with a sheet of paper or card and pull it down until you come to the line at the end of the frame. Read the frame and write down your answer to the question. This is most important – your learning will be much greater if you commit yourself actively by writing your answer down. You can check immediately whether or not your answer is right because each frame starts with the correct answer to the previous frame's question.

If you work through the whole programme in this way you will be learning at your own pace and checking on your progress as you go. If you are working at about the right pace you should get most of the questions right, but if you get one wrong you should read the frame again, look at the question, its answer, and any explanation offered, and try to understand how the answer was obtained. When you are satisfied about the answer go on to the next frame.

Learning a subject (as opposed to just reading a book about it) can be a long job. Don't get discouraged if you find the programmes taking a long time. Some students find this subject easy and work through each programme in about an hour or even less. Others have been known to take up to four hours for some programmes. Provided the programme objectives are achieved the time spent is relatively unimportant.

After completing each programme try the test at the end and only proceed to the next programme if your test score is up to the standard indicated.

Each programme finishes with a page of revision notes which should be helpful either to summarise the programme before or after use, or to serve as revision material later.

I hope you find the programmes enjoyable and useful.

Symmetry Elements and Operations

Objectives

After completing this programme, you should be able to:

1. Recognise symmetry elements in a molecule.
2. List the symmetry operations generated by each element.
3. Combine together two operations to find the equivalent single operation.

All three objectives are tested at the end of the programme.

Assumed Knowledge

Some knowledge of the shapes of simple molecules is assumed.

Symmetry Elements and Operations

1.1   The idea of symmetry is a familiar one, we speak of a shape
       as being "symmetrical", "unsymmetrical" or even "more
       symmetrical than some other shape". For scientific purposes,
       however, we need to specify ideas of symmetry in a more
       quantitative way.

       Which of the following shapes would you call the more sym-
       metrical?

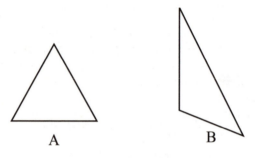

1.2   If you said A, it shows that our minds are at least working
       along similar lines!

       We can put the idea of symmetry on a more quantitative
       basis. If we rotate a piece of cardboard shaped like A by
       one third of a turn, the result looks the same as the starting
       point:

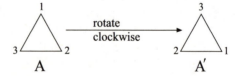

       Since A and A′ are *indistinguishable* (not identical) we say
       that the rotation is a symmetry operation of the shape.

       Can you think of another operation you could perform on a
       triangle of cardboard which is also a symmetry operation?
       (Not the anticlockwise rotation!)

1.3   Rotate by half a turn about an axis through a vertex i.e. turn it over

How many operations of this type are possible?

1.4   Three, one through each vertex.

We have now specified the first of our symmetry operations, called a PROPER ROTATION, and given the symbol C. The symbol is given a subscript to indicate the ORDER of the rotation. One third of a turn is called C_3, one half a turn C_2, etc.

What is the symbol for the operation:

1.5 C_4. It is rotation by $\frac{1}{4}$ of a turn.

A symmetry *operation* is the operation of actually doing something to a shape so that the result is indistinguishable from the initial state. Even if we do not do anything, however, the shape still possesses an abstract geometrical property which we term a symmetry *element*. The element is a geometrical property which is said to generate the operation. The element has the same symbol as the operation.

What obvious symmetry element is possessed by a regular six-sided shape:

1.6 C_6, a six-fold rotation axis, because we can rotate it by $\frac{1}{6}$ of a turn

One element of symmetry may generate more than one operation e.g. a C_3 axis generates two operations called C_3 and C_3^2:

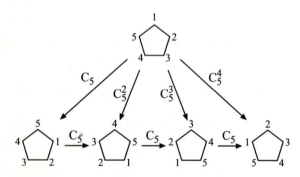

What operations are generated by a C_5 axis?

_____

1.7 C_5, C_5^2, C_5^3, C_5^4

What happens if we go one stage further i.e. C_5^5?

_____

1.8 We get back to where we started i.e.

The shape is now more than indistinguishable, it is IDENTICAL with the starting point. We say that C_5^5, or indeed any $C_n^n = E$, where E is the IDENTITY OPERATION, or the operation of doing nothing. Clearly this operation can be performed on anything because everything looks the same after doing nothing to it! If this sounds a bit trivial I apologise, but it is necessary to include the identity in the description of a molecule's symmetry in order to be able to apply the theory of Groups.

We have now seen two symmetry elements, the identity, E, and a proper rotation axis C_n. Can you think of a symmetry element which is possessed by all *planar* shapes?

1.9 A plane of symmetry.

This is given the symbol σ (sigma). The element generates only one operation, that of reflection in the plane.

Why only one operation? Why can't we do it twice – what is σ^2?

1.10 $\sigma^2 = E$, the identity, because reflection in a plane, followed by reflection back again, returns all points to the position from which they started, i.e. to the *identical* position.

Many molecules have one or more planes of symmetry. A flat molecule will always have a plane in the molecular plane e.g. H_2O, but this molecule also has one other plane.

Can you see where it is?

AT THIS STAGE SOME READERS MAY NEED TO MAKE
USE OF A KIT OF MOLECULAR MODELS OR SOME SORT
OF 3-DIMENSIONAL AID. IN THE ABSENCE OF A PROPER
KIT, MATCHSTICKS AND PLASTICINE ARE QUITE GOOD,
AND A FEW LINES PENCILLED ON A BLOCK OF WOOD
HAVE BEEN USED.

1.10a   You were trying to find a second
         plane of symmetry in the water
         molecule:

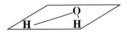

1.11   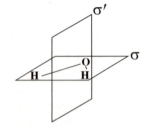 σ is the plane of the molecule,
                                σ' is at right angles to it and
                                reflects one H atom to the
                                other.

The water molecule can also be brought to an indistinguishable configuration by a simple rotation. Can you see where
the proper rotation axis is, and what its order is?

1.12 C_2, a twofold rotation axis, or rotation by half a turn.

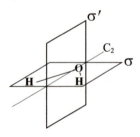

A C_2 axis passing through space is the hardest of all symmetry elements to see. It will be much easier to visualise if
you use a model of the molecule.

This completes the description of the symmetry of water. It
actually has FOUR elements of symmetry – one of which is
possessed by all molecules irrespective of shape. Can you list
all four symmetry elements of the water molecule?

1.13   E C_2 σ σ' Don't forget E!

Each of these elements generates only one operation, so the four symbols also describe the four operations.

Pyridine is another flat molecule like water. List its symmetry elements.

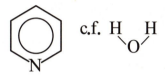

1.14   E C_2 σ σ' i.e. the same as water.

Many molecules have this set of symmetry elements, so it is convenient to classify them all under one name, the set of symmetry operations is called the C_{2v} point group, but more about this nomenclature later.

There is a simple restriction on planes of symmetry which is rather obvious but can sometimes be helpful in finding planes. A plane must either pass through an atom, or else that type of atom must occur in pairs, symmetrically either side of the plane. Take the molecule $SOCl_2$, which has a plane, and apply this consideration. Where must the plane be?

1.15   Through the atoms S and O because there is only one of each:

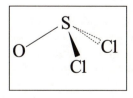

The molecule NH_3 possesses planes. Where must they lie?

1.16   Through the nitrogen (only one N), and through at least one
        hydrogen (because there is an odd number of hydrogens).
        Look at a model and convince yourself that this is the case.

        A further element of symmetry is the INVERSION CENTRE,
        i. This generates the operation of inversion through the centre.
        Draw a line from any point to the centre of the molecule, and
        produce it an equal distance the other side. If it comes to an
        equivalent point, the operation of inversion is a symmetry
        operation. e.g. ethane in the staggered conformation:

        N.B. The operation of
        inversion cannot be
        physically carried
        out on a model.

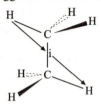

        Which of the following have inversion centres

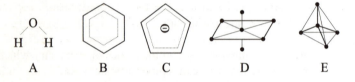

        A   B   C   D   E

1.17   Only B and D e.g., for C, the operation i would take point x
        to point y which is certainly not equivalent:

        An inversion centre may be:
        a. In space in the centre of a molecule (ethane, benzene); or
        b. At a single atom in the centre of the molecule (D above).

        If it is in space, all atoms must be present in even numbers,
        spaced either side of the centre. If it is at an atom, then that
        type of atom *only* must be present in an odd number. Hence a
        molecule AB_3 cannot have an inversion centre but a molecule
        AB_4 might possibly have one.

        Use this consideration to decide which of the following
        MIGHT POSSIBLY have a centre of inversion.

        NH_3 CH_4 C_2H_2 C_2H_4 $SOCl_2$ SO_2Cl_2

1.18 CH_4, C_2H_2, C_2H_4, SO_2Cl_2 fulfil the rules, i.e. have no atoms present in odd numbers, or have only one such atom.

Which of these actually have inversion centres?

1.19   Only C_2H_2 and C_2H_4. Both have an inversion centre midway between the two carbon atoms.

What is the operation i^2?

1.20 $i^2 = E$, for the same reason that $\sigma^2 = E$ (Frame 1.10).

We now have the operations E, σ, C_n, i. Only one more is necessary in order to specify molecular symmetry completely. That is called an **IMPROPER ROTATION** and is given the symbol S, again with a subscript showing the order of the axis. The element is sometimes called a rotation-reflection axis, and this describes the operation very well.

The S_n operation is rotation by $1/n$ of a turn, followed by reflection in a plane *perpendicular to the axis*, e.g. ethane in the staggered conformation has an S_6 axis because it is brought to an indistinguishable arrangement by a rotation of $1/6$ of a turn, followed by reflection:

N.B. Neither C_6 nor σ are present on their own.

In this example the effect of the symmetry operation has been shown by labelling one corner of the drawing. Draw the position of the label after the S_6 operation is applied a second time.

1.21

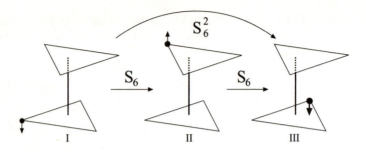

Now consider what single symmetry operation will take this molecule from state I direct to state III i.e. what single operation is the same as S_6^2?

1.22 $S_6^2 = C_3$, rotation by one third of a turn, because the molecule has been rotated by 2/6 of a turn ($= C_3$) and reflected twice ($\sigma^2 = E$).

What happens to the marker if S_6 is applied once more, i.e. what single operation has the same effect as S_6^3 (use a model or the diagram above).

1.23 $S_6^3 = i$. In general $S_n^{n/2} = i$ if n is even and n/2 is odd. The operation $S_n^{n/2}$ is then not counted by convention. If S_n (n even) is present, and n/2 is odd, i is present but the converse is not necessarily true.

Now apply S_6 once more, so that it has been applied four times in all.

What other operation gives the same result as S_6^4?

1.24 $S_6^4 = C_3^2$ for the same reason that $S_6^2 = C_3$ (Frame 1.22) i.e. we have now rotated by $1/6$ of a turn 4 times ($= C_3^2$), and reflected 4 times ($= E$)

S_6^5 is a unique operation, and $S_6^6 = E$. This is again true for any S_n of even n.

Let us now look at S_n of odd n because the case is rather different from even n. It may at first seem rather a trivial operation, because the C_n axis and a perpendicular plane must both be present, but it is necessary to include it to apply Group Theory to symmetry.

Use as the model a flat equilateral triangle with one vertex "labelled"; this label is only to help us to follow the effect of the operations, for example the application of S_3 moves the label as shown:

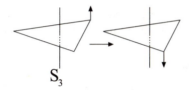

Write down the result of applying S_3 clockwise once, twice and then three times.

S_3

1.25

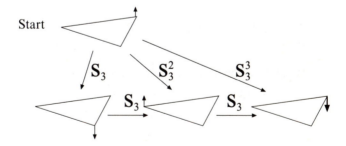

Start

S_3 S_3^2 S_3^3

S_3 S_3

In contrast to S_6 and C_3, applying the operation n times, where n is the order of the axis does not bring us back to the identity.

Keep going, then, when do we get E?

1.26

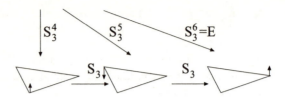

This result is quite general, for n odd $S_n^{2n} = E$, because we have rotated through two whole circles, and reflected an even number of times.

The equilateral triangle also has E, C_3, and σ among its elements of symmetry. Many of the operations we have generated by using the S_3 element of symmetry could have been generated by using other elements e.g., $S_3^2 = C_3^2$. Write these equivalents underneath the symbol S_3^n where appropriate:

S_3 S_3^2 S_3^3 S_3^4 S_3^5 S_3^6

e.g. C_3^2

1.27 S_3 S_3^2 S_3^3 S_3^4 S_3^5 S_3^6

           C_3^2 σ C_3 E

By convention, only S_3 and S_3^5 are counted as distinct operations generated by the S_3 symmetry element.

Do a similar analysis for the symmetry element C_6 (proper rotation axis) of benzene, which also has C_3 and C_2 axes colinear with the C_6. Clearly $C_6^2 = C_3$ since rotation by two sixths of a turn is the same as rotation by one third of a turn. Write the operations which have the same effect as C_6 C_6^2 C_6^3 C_6^4 C_6^5 and C_6^6.

1.28 C_6 C_6^2 C_6^3 C_6^4 C_6^5 C_6^6
       C_3 C_2 C_3^2 E

Again, by convention, only the operations C_6 and C_6^5 are counted, the others are taken to be generated by C_3 and C_2 axes colinear with C_6.

We have just been looking at the operations generated by a particular symmetry element, let us now turn to the identification of symmetry elements in a molecule. You must first be quite sure you appreciate the difference between a symmetry *element* and the symmetry *operation(s)* generated by the element. If you are not confident of this point, have another look at frames 1.5 to 1.13.

Some molecules have a great many symmetry elements, some of which are not immediately obvious e.g. XeF_4:

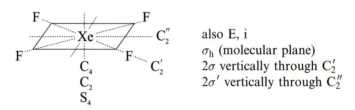

also E, i
σ_h (molecular plane)
2σ vertically through C_2'
$2\sigma'$ vertically through C_2''

Hence the complete list of symmetry elements is:

E C_4 C_2 S_4 i $2C_2'$ $2C_2''$ σ_h 2σ $2\sigma'$

List the symmetry elements of the following molecules:

(assume CH_3 groups spherical)

If there is a set of, say, three equivalent planes, write them as 3σ, but if there are three non-equivalent planes, write $\sigma\,\sigma'\,\sigma''$. Similarly for other elements.

1.29   BCl$_3$:   E   C$_3$ S$_3$ 3C$_2$ 3σ σ (a somewhat similar

         NH$_3$:   E   C$_3$ 3σ case to XeF$_4$)

         Butene: E   C$_2$ σ i

We will now look at what happens if two symmetry operations are combined, or performed one after the other. The result is always the same as doing one symmetry operation alone, so we can write an equation such as:

$$\sigma C_2 = \sigma'$$

This equation means that the operation C$_2$ *followed* by the operation σ gives the same result as the operation σ'. Note that the order in which the operations are performed is from right to left. I apologise for the introduction of back to front methods, but this is the convention universally used in the mathematics of operators, and the reason for it will become evident when we begin to use matrices to represent symmetry operations.

Confirm that this relationship is in fact true for the water molecule. It may help to put a small label on your model to show the effect of applying the operations:

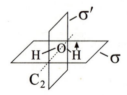

Draw the position of the arrow after applying C$_2$, and then after applying σ to the result. Hence confirm that $\sigma C_2 = \sigma'$.

1.30

What is the effect of reversing the order of the operations? i.e. what is the product C$_2\sigma$ (σ followed by C$_2$)?

1.31

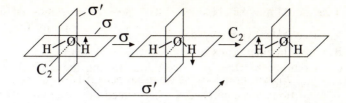

In this case the two operations COMMUTE i.e., $\sigma C_2 = C_2\sigma$, but this is not always true.

Use this diagram with an arrow to set up a complete multiplication table for the symmetry OPERATIONS of the water molecule, putting the product of the top operation, then the side operation, in the spaces:

	E	C_2	σ	σ'
E				
C_2				
σ				
σ'				

1.32

	E	C_2	σ	σ'
E	E	C_2	σ	σ'
C_2	C_2	E	σ'	σ
σ	σ	σ'	E	C_2
σ'	σ'	σ	C_2	E

You should now be able to:

A.   Recognise symmetry elements in a molecule.
B.   List the operations generated by each element.
C.   Combine together two operations to find the equivalent single operation.

I'm afraid the next page is a short test to see how well you have learned about elements and operations. After you have done it, mark it yourself, and it will give you some indication of how well you have understood this work.

Symmetry Elements and Operations Test

1.   List the symmetry elements of the molecules.

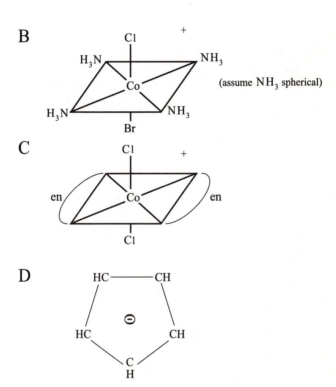

A CH_3 CH_3

$C = C$ (assume CH_3 spherical)

H   H

B   Cl   +

H_3N NH_3

Co

H_3N NH_3

Br   (assume NH_3 spherical)

C   Cl   +

en   Co   en

Cl

D   HC———CH

HC   ⊖   CH

C
H

2.   Set up the multiplication table for the *operations* of the mole-
      cule *trans* but-2-ene. Apply the top operation then the side
      operation:

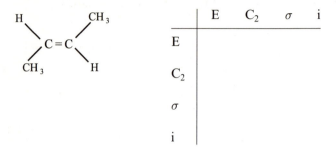

	E	C_2	σ	i
E				
C_2				
σ				
i				

3.   In this question you have to state the single symmetry opera-
      tion of XeF_4 which has the same effect as applying a given
      operation several times. The diagram below shows the location
      of the symmetry elements concerned.

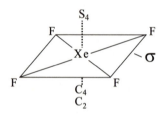

What operation has the same effect as:

A. S_4^2 E. C_4^3

B. S_4^3 F. C_4^4

C. S_4^4 G. σ^2

D. C_4^2 H. i^2

Answers

Give yourself one mark for each underlined answer you get right. (The others are so easy, they are not worth a mark!)

1.   A. $\underline{E}$ $\underline{C_2}$ $\underline{\sigma}$ $\underline{\sigma'}$

    B.   E $\underline{C_4}$ $\underline{C_2}$ $\underline{2\sigma}$ $\underline{2\sigma'}$

    C.   E $\underline{C_2}$ $\underline{C_2}$ $\underline{C_2}$ $\underline{i}$ $\underline{\sigma}$ $\underline{\sigma'}$ $\underline{\sigma''}$

    D.   E $\underline{C_5}$ $\underline{5C_2}$ $\underline{\sigma}$ $\underline{5\sigma'}$ $\underline{S_5}$

Total = 20

2.

	E	C_2	σ	i
E	E	C_2	σ	i
C_2	C_2	$\underline{E}$	$\underline{i}$	$\underline{\sigma}$
σ	σ	$\underline{i}$	$\underline{E}$	$\underline{C_2}$
i	i	$\underline{\sigma}$	$\underline{C_2}$	$\underline{E}$

Total = 9

3.   A. $S_4^2 = \underline{C_2}$ E. $C_4^3 = \underline{C_4^3}$

    B. $S_4^3 = \underline{S_4^3}$ F. $C_4^4 = \underline{E}$

    C. $S_4^4 = \underline{E}$ G. $\sigma^2 = \underline{E}$

    D. $C_4^2 = \underline{C_2}$ H. $i^2 = \underline{E}$

Total = 8

Grand Total = 37

To be able to proceed confidently to the next programme you should
have obtained at least:

> Question 1 (Objective 1) 15/20 (Frames 1.1–1.20)
> Question 2 (Objective 2) 7/9   (Frames 1.28–1.32)
> Question 3 (Objective 3) 4/8   (Frames 1.6–1.10, 1.19–1.28).

If you have not obtained these scores you would be well
advised to return to the frames shown, although a low
score on question 3 is less serious than the other two.

Symmetry Elements and Operations

Revision Notes

The symmetry of a molecule can be described by listing all the symmetry elements of the molecule. A molecule possesses a symmetry element if the application of the operation generated by the element leaves the molecule in an *indistinguishable* state. There are five different elements necessary to completely specify the symmetry of all possible molecules:

E   the identity

C_n proper rotation axis of order n

σ a plane of symmetry

i   an inversion centre

S_n improper (or rotation-reflection) axis of order n.

Each of the elements E, σ, i only generates one operation, but C_n and S_n can generate a number of operations because the effect of applying the operation a number of times can count as separate operations e.g., the C_3 element generates operations C_3 and C_3^2. Some such multiple applications of an operation have the same effect as a single application of a different operation. In these cases only the single case is counted, e.g., $C_4^2 = C_2$, and only C_2 is counted.

If two operations are performed successively on a molecule, the result is always the same as the application of only one different operation. It is therefore possible to set up a multiplication table for the symmetry operations of a molecule to show how the operations combine together.

When writing an equation to represent the successive application of symmetry elements it is necessary to remember that $\sigma \, \sigma' \, C_4$ means C_4 followed by σ', followed by σ.

Point Groups

Objectives

After completing this programme you should be able to:

1. State the point group to which a molecule belongs.
2. Confirm that the complete set of symmetry operations of a molecule constitutes a group.
3. Arrange a set of symmetry operations into classes.

The first of these objectives is vital to the use of group theory and is the only one tested at the end of the programme.

Assumed Knowledge

A knowledge of simple molecular shapes, and of the contents of Programme 1 is assumed.

Point Groups

2.1   Write down the symbols of the FIVE elements needed to completely specify molecular symmetry.

2.2   E   C   S σ i

What are the names of these five elements of symmetry?

2.3   E — The identity element
C — Proper rotation axis
S — Improper rotation (or rotation-reflection) axis
σ — Plane of symmetry
i — Inversion centre

List all the symmetry elements of

2.4   E C_3 $3C_2$ σ $3\sigma'$ S_3

If you have got these three questions substantially correct you may proceed, otherwise return to Programme 1 – Symmetry Elements and Operations.

List all the symmetry elements of

2.5   E C_3 $3C_2$ σ $3\sigma'$ S_3

i.e. exactly the same as BCl_3

There are many other examples of several molecules having the same set of symmetry elements, e.g. list all the symmetry elements of

---

2.6   All three of these molecules (and many more!) have the elements

E C_2 σ σ'

In the same way all square planar molecules contain the elements E C_4 $C_2(= C_4^2)$ $4C_2$ σ $4\sigma'$ i S_4, regardless of the chemical composition of the molecule e.g.

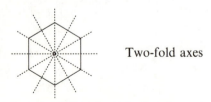

etc.

It is convenient to classify all such molecules by a single symbol which summarises their symmetry. This symbol for a flat square molecule is D_{4h}.

Can you suggest the symbol for a flat hexagonal molecule like benzene:

2.7 D_{6h} the symmetry is similar to that of the square planar case, but the principal axis is a 6-fold axis not a 4-fold axis.

The symmetry symbol consists of three parts:

The number indicates the order of the principal (i.e. highest order) axis. This is conventionally taken to be vertical.

The small letter h indicates a horizontal plane.

The capital letter D indicates that there are n(= 6 for benzene) C_2 axes at right angles to the principal C_n axis (C_6 for benzene):

Two-fold axes

How many two-fold axes like this are there in a flat square molecule like XeF_4?

2.8   Four

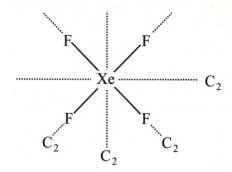

Let us look now at a flat triangular molecule, say BCl_3:

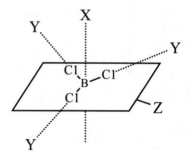

What are the symmetry elements labelled X, Y, and Z?

2.9   X = C_3 axis
         Y = C_2 axes
         Z = plane of symmetry

The principal C_3 axis is taken, conventionally to be vertical, so the plane is a horizontal plane (σ_h), and there are three C_2 axes at right angles to the principal axis.

What, therefore, is the symmetry symbol of the BCl_3 molecule? (frame 2.7 may help).

2.10 D_{3h}

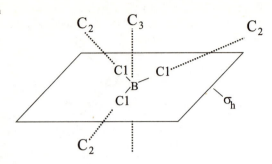

Point group symbol: D_{3h}

3C_2 axes   3-fold principal   horizontal plane
(horizontal)   axis (vertical)

The molecule is said to belong to the D_{3h} POINT GROUP.

Let us now get a bit more general, and call the principal axis C_n, so that its order, n, can be any number.

If there is no horizontal plane of symmetry, but there are n vertical planes as well as nC_2 axes, the point group is D_{nd}.

The D and the number mean the same as before but the small d stands for DIHEDRAL PLANES, because the n vertical planes lie between the nC_2 axes.

Ethane in the staggered conformation belongs to a D_{nd} point group. Decide on the value of n from the following diagram (looking down the principal axis), and hence state the point group to which ethane belongs.

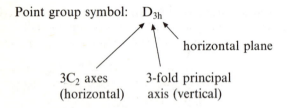

2.11 D_{3d}, a model will help to convince you of the elements of symmetry in this case, but the following diagram is looking down the principal, vertical, 3-fold axis:

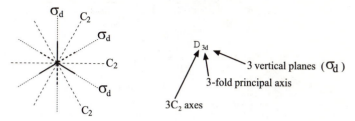

This is another case like frame 1.12 in which the C_2 axis passes through space and not along a bond. These axes are quite difficult to see and a molecular model may be necessary.

In the eclipsed conformation ethane has an additional element of symmetry. Can you see from the diagram (or a model) what the extra element is?

2.12   A horizontal plane of symmetry, σ_h

What does this make the point group of ethane in the eclipsed conformation?

$$H\text{—}C\begin{matrix}H\\H\end{matrix}$$

2.13 D_{3h} i.e. in the eclipsed conformation the horizontal plane takes precedence over the dihedral planes in describing the symmetry.

Some molecules have a principal C_n axis, and nC_2 axes at right angles, but no horizontal or vertical (dihedral) planes.

There is then no need to include h or d in the symmetry symbol. If the principal axis is a 3-fold axis what is the symmetry symbol in this case?

2.14 D_3 i.e., it has a 3-fold axis and three C_2 axes at right angles, hence D_3, but no σ_h or a σ_d, so no additional symbol is necessary.

An example of an ion of this symmetry is:

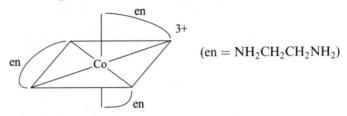

(en $= NH_2CH_2CH_2NH_2$)

You will probably need a model of the ion to see the axes, although an alternative diagram of the structure shows its symmetry very well:

If the principal C_n axis is not accompanied by nC_2 axes, the first letter of the point group is C. A horizontal plane is looked for first, and is shown by a little h. If σ_h is not present, n vertical planes are looked for and are shown by a small v.

e.g.

$$O$$
$$H \quad H$$
$$C_2$$

C_2, no C_2 at right angles no σ_h, but $2\sigma_v$ ∴ point group C_{2v}

$$N$$
$$H$$
$$H \quad H$$

What is the point group of

2.15 C_{3v} i.e. it has a principal C_3 axis and 3 vertical planes.

Remember that all flat molecules have a plane of symmetry in the molecular plane. Try to decide the point group of a free boric acid molecule which has no vertical planes or horizontal C_2 axes.

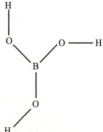

2.16 C_{3h} i.e. it has a principal C_3 axis, no horizontal C_2 axes, and
        a horizontal plane

        What is the point group of the flat ion:

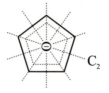

2.17 D_{5h} i.e. it has a C_5 (vertical), 5 C_2 axes at right angles, and a
        horizontal plane.

        List the four symmetry
        elements of fumaric acid:
        (CARE! There is again a
        C_2 axis through space).

H⎯⎯⎯⎯COOH

C = C

HOOC⎯⎯⎯⎯H

2.18   E, C_2, σ_h, i. What does this make the point group symbol?

2.19 C_{2h} i.e. it has a C_2 axis and a horizontal plane.

        The molecule H_2O_2 and the ion cis[Co(en)$_2$Cl$_2$]$^+$ both have
        only the identity and one proper axis of symmetry. They both
        belong to the same point group. Can you say which one it is?

        (A model, or the diagrams below, might help.)

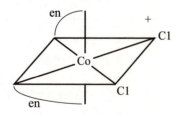

2.20 C_2. They both have a C_2 axis:

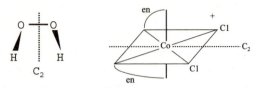

We have so far seen the point groups, D_{nh}, D_{nd} D_n, C_{nh}, C_{nv} and C_n. These groups cover many real molecules, even simple linear ones which have an infinity-fold axis e.g.

$$\begin{array}{cc} \begin{array}{c} H \\ | \\ C \\ ||| \\ C \\ | \\ H \end{array} D_{\infty h} & \begin{array}{c} H \\ | \\ Cl \end{array} C_{\infty v} \end{array}$$

There are three additional groups for highly symmetrical molecules, octahedral molecules belong to the group O_h, tetrahedral molecules to T_d, and icosahedral structures to I_h. You must realise that T_d refers to the symmetry of the whole molecule e.g. CH_4 and CCl_4 both belong to the T_d group, but $CHCl_3$ does not.

What is the point group of $CHCl_3$?

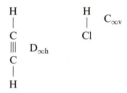

2.21 C_{3v}
Some rather rare molecules possess only two elements of symmetry, and these are given a special symbol:

E and i only C_i
E and σ only C_s
E and S_n only S_n

Many molecules have no symmetry at all (i.e. their only symmetry element is the identity, E. Such molecules belong to the C_1 point group.

The following are examples of molecules with only one or two symmetry elements.

What are their point groups?

A.
$$\begin{array}{c} Cl \\ | \\ C \\ F \diagup \quad \diagdown H \\ H \end{array}$$

B.
$$\begin{array}{c} F \\ | \\ C \\ Br \diagup \quad \diagdown H \\ Cl \end{array}$$

2.22   A. C_S

        B. C_1

There is a simple way of classifying a molecule into its point group, and a sheet at the end of this programme gives this. You will see that the tests at the bottom of the scheme are similar to those used to introduce the nomenclature in this programme. The scheme does not test for all the symmetry elements of a molecule, only certain key ones which enable the point group to be found unambiguously.

Have a look at the sheet, and try to follow it through for the ion:

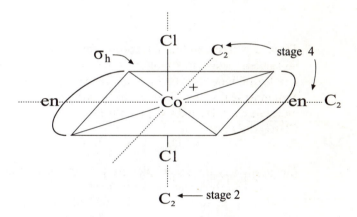

Stage 1 – it is not one of these special groups
Stage 2 – there is a C_2 axis – ∴ n = 2
Stage 3 – there is no S_4 colinear with C_2
Stage 4 – there are two C_2 axes at right angles, there is a horizontal plane.

What point group have you arrived at? (Remember the value of n found in Stage 2.)

2.23 D_{2h}

Use the scheme to find the point group of each of the following: (C, E, F and G are a bit tricky without a model, but
you may get C, F and G right by analogy with ethane as
discussed in frames 2.10–2.13).

A CH_3 H   B   H   H   C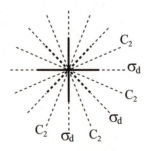
      \\   /   \\   /
       C = C   C = C
      /   \\   /   \\
     H CH_3 CH_3 CH_3

D   S   E   H—N
        ‖   N   H
     / ╲—Cl |
    O   Cl   N
                                | ╲
                                H   H

F   ⬠⬠ Ru   G   ⬠
                              Fe
                            ⬠

2.24   A. C_{2h} B. C_{2v} C. D_{4d} D. C_s E. C_2 F. D_{5h} G. D_{5d}

The hardest of these examples are probably C and G which
are both D_{nd} molecules. It is often very difficult to see the n
2-fold axes on such a molecule and you may need to ask
advice on this. Frame 2.11 shows the axes in the case of a
D_{3d} molecule. The corresponding diagram, looking down the
principal 4-fold axis of $Mn_2(CO)_{10}$ is:

C_2
σ_d
C_2
σ_d
C_2' σ_d C_2

A simple rule to remember is that any n-fold staggered structure (like C_2H_6, $Mn_2(CO)_{10}$ etc) belongs to the point group D_{nd}, and you may find it easier simply to remember this rule.

We have said that the symbol represents the POINT GROUP of the molecule. This is because all the symmetry elements of a molecule always pass through one common point (sometimes through a line or a plane, but always through a point).

Where is the point for examples A and G above?

2.25   A – the centre of the $C\!=\!C$ double bond
G – the Fe atom

At this stage, the programme begins to look at what mathematicians call a GROUP. If you have had enough for one sitting, this is a convenient place to stop, but in any case it is not absolutely vital for a chemist to know about the rules defining a group, although I strongly recommend you to work through the rest of the programme. You should now be able to classify a molecule into its point group, which is absolutely vital to the use of Group Theory, and the test at the end of the programme tests only this classification.

The term GROUP has a precise mathematical meaning, and the set of symmetry OPERATIONS of a molecule constitutes a mathematical group. A group consists of a set of members which obey four rules:

a.   The product of two members, and the square of any member is also a member of the group.
b.   There must be an identity element.
c.   Combination must be associative i.e. $(AB)C = A(BC)$.
d.   Every member must have an inverse which is also a member i.e. $AA^{-1} = E$, the identity, if A is a member, A^{-1} must also be.

N.B. Some texts use the word *element* for the members of a group. This convention has not been followed here in order to avoid confusion with the term *symmetry element*. It is the set of *symmetry operations* which form the group.

Let us take the C_{2v} group (e.g. H_2O) and confirm these rules. The group has four operations, E, C_2, σ, σ':

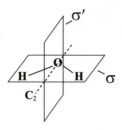

We have already seen the effect of combining two operations in the programme on elements and operations.

Set up the complete multiplication table for the group operations (in Programme 1 you used a little arrow on H to help do this).

	E	C_2	σ	σ'
E				
C_2				
σ				
σ'				

2.26

	E	C_2	σ	σ'
E	E	C_2	σ	σ'
C_2	C_2	E	σ'	σ
σ	σ	σ'	E	C_2
σ'	σ'	σ	C_2	E

If you did not get this result, look back at the first pro-gramme, frames 1.29–1.32.

We can see immediately from this table that rules a and b are true for this set of operations.

What about rule d? What is the inverse of σ', i.e. what multi-plies with σ' to give E?

2.27 σ', it is its own inverse, $\sigma'\sigma' = E$. This is true for all the operations of this group.

Consider the C_3 element in a D_{3h} molecule. What is the inverse of the C_3 operation, or what *operation* will bring the shape back to the starting point (I'd rather you didn't say C_3 in the opposite direction!).

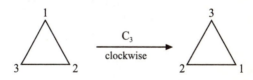

2.28 C_3^2, i.e. apply the C_3 operation clockwise a further two times. Thus $C_3^2 C_3 = C_3^3 = E$. (Remember that this means C_3 fol-lowed by C_3^2.)

Note particularly that it is the symmetry OPERATIONS, not the elements which form a group.

Confirm rule c for the elements C_2, σ, and σ' of the C_{2v} group, i.e. work out the effect of $(C_2\sigma)\,\sigma'$ and of $C_2\,(\sigma\sigma')$.

2.29 $(C_2\sigma)\sigma' = \sigma'\sigma' = E$
       $C_2(\sigma\sigma') = C_2C_2 = E$

i.e. the operations are associative.

The C_{2v} point group only has four operations, so it is a simple matter to set up the group multiplication table. There is, however, a further feature of groups which can only be demonstrated by using a rather larger group such as C_{3v}. Ammonia belongs to the C_{3v} group. Can you write down the five symmetry *elements* of ammonia?

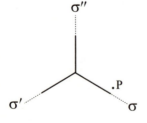

2.30   E C_3 3σ

What operations do these elements generate?

2.31   E C_3 C_3^2 σ σ' σ'' (or 3σ)

We can set up the 6×6 multiplication table for these operations by considering the effect of each operation on a point such as P in the diagram below, which has the C_3 axis perpendicular to the paper:

The C_3 and C_3^2 operations are clockwise

Draw the position of point P after applying C_3 and then σ' (call the new position P').

2.32

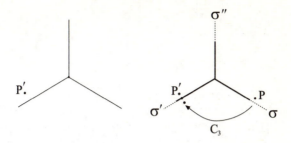

What single operation would take P to P′?

2.33 σ''

i.e. $\sigma' C_3 = \sigma''$ (remember that this means C_3 followed by σ' has the same effect as σ'' — we write the operations in reverse order).

What happens if we do it the other way round, i.e. what is σ' followed by C_3 (= $C_3 \sigma'$)?

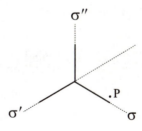

2.34 σ

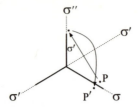

In this case $\sigma'C_3$ does not equal $C_3\sigma'$ – we say that these two operations do not COMMUTE.

Use the effect of the group operations on the point P to see which of the following pairs of operations commute:

C_3 and C_3^2 σ and C_3 σ and σ' E and C_3^2

2.35 $C_3C_3^2 = E;$ $C_3^2C_3 = E$ i.e. C_3 and C_3^2 commute
        $\sigma C_3 = \sigma';$ $C_3\sigma = \sigma''$ i.e. σ and C_3 do not commute
        $\sigma\sigma' = C_3;$ $\sigma'\sigma = C_3^2$ i.e. σ and σ' do not commute
        $EC_3^2 = C_3^2;$ $C_3^2E = C_3^2$ i.e. E and C_3^2 commute

It should be obvious that E commutes with everything – it does not matter if you do nothing before or after the operation!

We will now consider briefly the subject of CLASSES of symmetry operations. Two operations A and B are in the same class if there is some operation X such that:

$XAX^{-1} = B$ (X^{-1} is the inverse of X, i.e. $XX^{-1} = E$)

We say that B is the *similarity transform* of A, and that A and B are *conjugate*.

Since any σ is its own inverse we can perform a similarity transformation on the operation C_3 by finding the single operation equivalent to $\sigma C_3 \sigma$.

Work out the position of point P after carrying out these three operations.

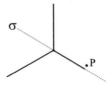

C_3 is clockwise

2.36

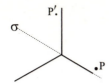

i.e.

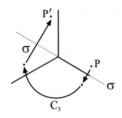

What single operation is the same as $\sigma C_3 \sigma$?

2.37 C_3^2. Thus C_3 and C_3^2 are in the same class.

What is the inverse of C_3?

2.38 C_3^2. Work out the similarity transform of σ by C_3, i.e. decide the operation equivalent to $C_3^2 \sigma C_3$.

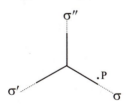

2.39 $C_3^2 \sigma C_3 = \sigma''$

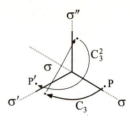

Thus σ and σ'' are in the same class

The complete set of symmetry operations of the C_{3v} point group, grouped by classes, is as follows:

E   (always in a class by itself)
C_3 C_3^2
σ σ' σ''

The operations are commonly written in classes as:

E $2C_3$ 3σ

It is not necessary to go through the whole procedure of working out similarity transformations in order to group operations into classes. A set of operations are in the same class if they are *equivalent operations* in the normally accepted sense. This is probably fairly evident for the example above.

The D_{3h} group (e.g. BCl_3) consists of the operations

E C_3 C_3^2 C_2 C_2' C_2'' σ_h S_3 S_3^5 σ_v σ_v' σ_v''

Group these operations into their six classes

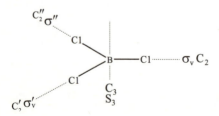

2.40   E

       $2C_3$

       $3C_2$

       σ_h

       $2S_3$

       $3\sigma_v$ (all equivalent but different from σ_h)

You should now be able to:

State the point group to which a molecule belongs.
Confirm that a set of operations constitutes a group.
Arrange a set of operations into classes.

The assignment of a molecule to its correct point group is a vital preliminary to the use of group theory, and this is the subject of the test which follows. The other two objectives of this programme are not tested because it is known in all cases that the symmetry operations of a molecule do constitute a group, and the tables (character tables) which are used in working out problems show the operations grouped by classes.

Point Groups Test

Classify the following molecules and ions into their point group. You may use molecular models and the scheme for classifying molecules.

1. CH_2Cl_2

2.

     H   H   H
      \\   /   \\   /
       B   B
    /   \\   /   \\
  H   H   H

3.   Cyclohexane (chair)
        (use a model)

4.   Cyclohexane (boat)

5. $O = P \overset{\textstyle Cl}{\underset{\textstyle Cl}{\rule[0.5ex]{2em}{0.4pt}Cl}}$

6. $F_4P F_2$ (trigonal bipyramidal arrangement of F around P)

11.	CBr_4
12.	SF_6
13.	CO_2
14.	OCS

7.   (imidazole, with N, N–H ring)

8.   Cr (staggered)

9. $[Cr(ox)_3]^{3-}$ ox = oxalate (a model is almost essential)

10. $Cr(H_2O)_2(ox)_2$ (a model is valuable)

Answers

One mark each.

1.	C_{2v}	8.	D_{6d}
2.	D_{2h}	9.	D_3
3.	D_{3d}	10.	C_2
4.	C_{2v}	11.	T_d
5.	C_{3v}	12.	O_h
6.	D_{3h}	13.	$D_{\infty h}$
7.	C_s	14.	$C_{\infty v}$

To be able to proceed confidently to the next programme you should have obtained at least 10 out of 14 on this test, and you should understand the assignment of the point group in any cases you got wrong.

If you are in any doubt about the assignment of point groups, return to frames 2.7 to 2.24.

Point Groups

Revision Notes

The set of symmetry operations of any geometrical shape forms a mathematical group, which obeys four rules:

i.   The product of two members of the group, and the square of any member, is also a member of the group.
ii. There must be an identity element.
iii. Combination must be associative, i.e. $(AB)C = A(BC)$
iv. Every member must have an inverse, i.e. if A is a member, then A^{-1} must also be a member, where $AA^{-1} = E$.

Symmetry operations do not necessarily commute, i.e. AB does not always equal BA.

A molecule can be assigned to its point group by a method which does not require the listing of all symmetry operations of the molecule; the method merely involves looking for certain key symmetry elements. The symbol for most molecular symmetry groups is in three parts e.g.

$$C_{4v} \quad C_{2h} \quad D_{3h} \quad D_{6d}$$

These have the following meanings:

i.   The number indicates the order of the principal (highest order) axis. This axis conventionally defines the vertical direction.
ii. The capital letter is D if an n-fold principal axis is accompanied by n two-fold axes at right angles to it; otherwise the letter is C.
iii. The small letter is h if a horizontal plane is present. If n vertical planes are present, the letter is v for a C group but d (= dihedral) for a D group. (N.B. h takes precedence over v or d.) If no vertical or horizontal planes are present, the small letter is omitted.

Systematic Classification of Molecules into Point Groups

C = rotation axis   i = inversion centre
S = improper axis (alternating axis) σ = plane of symmetry

1.   Examine for special groups

   a.   Linear, no σ perpendicular to molecular axis — $C_{\infty v}$
   b.   Linear, σ perpendicular to molecular axis — $D_{\infty h}$
   c.   Tetrahedral — T_d
   d.   Octahedral — O_h
   e.   Dodecahedral or icosahedral — I_h

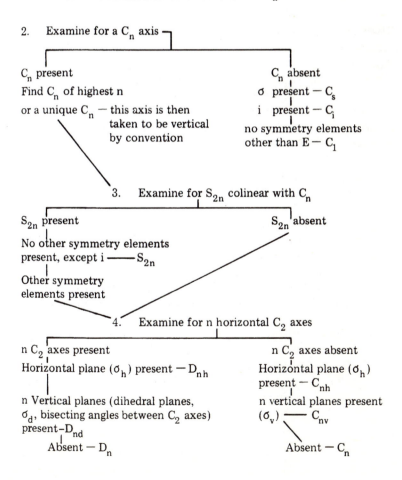

2.   Examine for a C_n axis

C_n present
Find C_n of highest n
or a unique C_n — this axis is then
                     taken to be vertical
                     by convention

C_n absent
σ present — C_s
i   present — C_i
no symmetry elements
other than E — C_1

3.   Examine for S_{2n} colinear with C_n

S_{2n} present
No other symmetry elements
present, except i —— S_{2n}

Other symmetry
elements present

S_{2n} absent

4.   Examine for n horizontal C_2 axes

n C_2 axes present
Horizontal plane (σ_h) present — D_{nh}

n Vertical planes (dihedral planes,
σ_d, bisecting angles between C_2 axes)
present – D_{nd}
     Absent — D_n

n C_2 axes absent
Horizontal plane (σ_h)
present — C_{nh}
n vertical planes present
(σ_v) —— C_{nv}

Absent — C_n

Non-degenerate Representations

Objectives

After completing this programme you should be able to:

1. Form a non-degenerate representation to describe the effect of the symmetry operations of a group on a direction such as x.

2. Reduce a reducible representation to its component irreducible representations.

Both objectives are tested at the end of the programme.

Assumed Knowledge

A knowledge of the shapes of p and d atomic orbitals, and of the contents of Programmes 1 and 2 is assumed.

Non-degenerate Representations

3.1   What are the point groups of the following molecules?

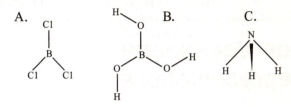

A.

B.

C.

3.2   A. D_{3h}
        B. C_{3h}
        C. C_{3v}

If you are quite happy about point groups, continue with this programme, if not, return to Programme 2 — Point Groups.

We are now going to progress one stage further in the quantitative description of molecular symmetry by using numbers to represent symmetry operations. These numbers are called REPRESENTATIONS (not unreasonably!), and in this programme we shall be mainly concerned with the numbers $+1$ and -1 so your maths should not be strained too far!

We shall initially use atomic p orbitals to illustrate the features of representations, but you must remember that the features we discover apply to many other directional properties as well.

Let us look at the effect on a p_x orbital of a C_2 rotation about the z axis:

The sign of the p_x orbital is changed, so how can the operation be represented, by $+1$ or -1?

3.3   −1. p_x becomes $-p_x$ or:
                                  $$C_2 p_x = -1 p_x$$

Let us look at the effect of various reflections on the p_x orbital — consider first a reflection in the xz plane which passes through the orbital:

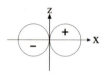

What does the orbital look like after applying the $\sigma(xz)$ operation?

3.4   Just the same, because the plane passes through the middle of both lobes.

What number will represent the operation $\sigma(xz)$?

3.5   +1   i.e. $\sigma(xz)\ p_x = 1 p_x$

What about the reflection in the yz plane — what is the result of $\sigma(yz)p_x$, and hence what number represents $\sigma(yz)$?

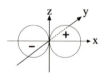

3.6 $\sigma(yz)p_x = -p_x$, hence $\sigma(yz)$ is represented by -1:

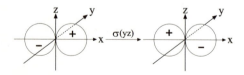

What number represents the effect of the identity operation, E?

3.7   + 1

We have now looked at the numbers representing the four operations E C_2 $\sigma(xz)$ $\sigma(yz)$.

These four operations form a group, can you remember which one it is?

3.8 C_{2v}

We say that the four numbers form the B_1 representation of the C_{2v} group:

C_{2v}	E	C_2	$\sigma(xz)$	$\sigma(yz)$	
B_1	1	-1	1	-1	x

Don't worry at this stage about the nomenclature B_1 — the symbol does carry information, but you can regard it simply as a label for the present.

We also say that x belongs to the B_1 representation of C_{2v} because this set of numbers represents the effect of the group operations on a p_x orbital, or indeed anything with the same symmetry properties as the x axis.

If our set of numbers represents the group operations, it should also represent the way the group operations combine together. Use a little arrow on the water molecule to find the product of the two operations C_2 and $\sigma(xz)$ like you did in an earlier programme:

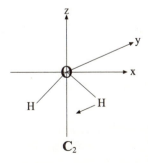

3.9 $C_2\sigma(xz) = \sigma(xz)C_2 = \sigma(yz)$

(See Programme 1 frames 1.29–1.31 if you did not get this result.)

Is this multiplication paralleled by the multiplication of the numbers representing the operations?

3.10   Yes $-1 \times 1 \quad = -1$
$\quad\quad\quad C_2 \times \sigma(xz) = \sigma(yz)$

The complete multiplication table for C_{2v} is:

C_{2v}	E	C_2	$\sigma(xz)$	$\sigma(yz)$
E	E	C_2	$\sigma(xz)$	$\sigma(yz)$
C_2	C_2	E	$\sigma(yz)$	$\sigma(xz)$
$\sigma(xz)$	$\sigma(xz)$	$\sigma(yz)$	E	C_2
$\sigma(yz)$	$\sigma(yz)$	$\sigma(xz)$	C_2	E

Write out the corresponding table for the numbers forming the B_1 representation.

3.11
B_1	1	-1	1	-1
1	1	-1	1	-1
-1	-1	1	-1	1
1	1	-1	1	-1
-1	-1	1	-1	1

Wherever C_2 or $\sigma(yz)$ appear in the first table, -1 appears in the second table, so the set of numbers is a genuine representation of the group.

Find the effect of the group operations on a p_y orbital, and hence derive a set of numbers which represent the effect of the operations on p_y.

3.12
$$E\,p_y = p_y$$
$$C_2\,p_y = -p_y$$
$$\sigma(xz)\,p_y = -p_y$$
$$\sigma(yz)\,p_y = p_y$$

E	is represented by		1
C_2	"	"	"	-1
$\sigma(xz)$	"	"	"	-1
$\sigma(yz)$	"	"	"	1

We say that y (or a p_y orbital) is SYMMETRIC to E and $\sigma(yz)$ and ANTISYMMETRIC to C_2 and $\sigma(xz)$ in C_{2v} symmetry. The p_y orbital thus belongs to the B_2 representation:

C_{2v}	E	C_2	$\sigma(xz)$	$\sigma(yz)$	
B_2	1	-1	-1	1	y

Set up the multiplication table for the B_2 representation, and confirm that it is a true representation (c.f. frame 3.11).

3.13

B_2	1	-1	-1	1
1	1	-1	-1	1
-1	-1	1	1	-1
-1	-1	1	1	-1
1	1	-1	-1	1

Wherever C_2 or $\sigma(xz)$ appear, there is -1.

Wherever E or $\sigma(yz)$ appear, there is 1.

The B_1 and B_2 representations are representations for two reasons:

i. The numbers represent the effect of the group operations on certain directional properties.
ii. The numbers multiply together in the same way as the group operations.

Find the representation of the C_{2v} point group to which a p_z orbital belongs, and confirm that the numbers multiply together in the same way as the operations:

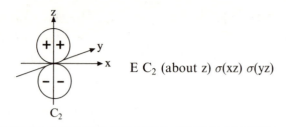

E C_2 (about z) $\sigma(xz)$ $\sigma(yz)$

3.14

C_{2v}	E	C_2	$\sigma(xz)$	$\sigma(yz)$	
A_1	1	1	1	1	z

The p_z orbital belongs to the **TOTALLY SYMMETRIC** or A_1 representation of the C_{2v} point group, because the p_z orbital is not changed by any of the group operations.

There is one further set of numbers called the A_2 representation which fulfills the two conditions given above for the C_{2v} point group. The full set of representations is included in a table called the **CHARACTER TABLE** of the group:

C_{2v}	E	C_2	$\sigma(xz)$	$\sigma(yz)$	
A_1	1	1	1	1	z
A_2	1	1	-1	-1	
B_1	1	-1	1	-1	x
B_2	1	-1	-1	1	y

The numbers in this table should strictly be called the **CHARACTERS** of the **IRREDUCIBLE REPRESENTATIONS** of the group. The meaning of this long title will become apparent in time.

Let us now try a slightly more complicated orbital, $3d_{xy}$. To which of the four representations of C_{2v} does this belong?

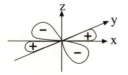

E C_2 (about z) $\sigma(xz)$ $\sigma(yz)$

3.15 A_2 E d_{xy} = d_{xy} representation   1

$C_2 d_{xy}$ = d_{xy} representation   1

$\sigma(xz) d_{xy}$ = $-d_{xy}$ representation -1

$\sigma(yz) d_{xy}$ = $-d_{xy}$ representation -1

It is also possible to find the representation to which other directional properties belong, e.g. a rotation about the x axis. If you hold a pencil horizontally in front of you and rotate it on its own axis (x), then, still rotating it, give it a half turn rotation about a vertical axis, its direction of rotation about its own axis will appear to have been reversed (try doing it!) Thus rotation about x is (symmetric/antisymmetric) to C_2.

3.16   Antisymmetric.

You need a particularly twisted mind to assign rotations to a symmetry class, and you may need to ask someone to explain it to you if you are not prepared to accept it.

The information we have just deduced is included in the full character table e.g.:

C_{2v}	E	C_2	$\sigma(xz)$	$\sigma(yz)$			
A_1	1	1	1	1	z		$x^2 - y^2, z^2$
A_2	1	1	-1	-1		R_z	xy
B_1	1	-1	1	-1	x	R_y	xz
B_2	1	-1	-1	1	y	R_x	yz

This shows the transformation properties of d orbitals as well as the x, y, and z directions and the three rotations about the x, y, and z axes called R_x, R_y and R_z. Some character tables may show even more — e.g. the representations to which f orbitals and polarisability components belong, but this is sufficient for our purposes now.

Is the set of numbers 3 3 1 1 a representation of C_{2v} in the sense we have been discussing representations? (Yes or no.)

3.17   No. Because $E \times E = E$ but $3 \times 3 = 9$ etc.

The numbers are, however, a set of CHARACTERS OF A REDUCIBLE REPRESENTATION of the C_{2v} group. Again, the meaning of this long title will become apparent later, but we may (rather loosely) abbreviate the title and call the set of numbers simply a REDUCIBLE REPRESENTATION.

The reducible representation 3 3 1 1 has been obtained simply by adding the representations $2A_1 + A_2$:

A_1	1	1	1	1
A_1	1	1	1	1
A_2	1	1	-1	-1
$2A_1 + A_2$	3	3	1	1

Can you see how the reducible representation 3 -1 -1 -1 is obtained?

3.18 $A_2 + B_1 + B_2$ i.e.

A_2	1	1	-1	-1
B_1	1	-1	1	-1
B_2	1	-1	-1	1
$A_2 + B_1 + B_2$	3	-1	-1	-1

We say that the reducible representation 3 -1 -1 -1 can be reduced to its component irreducible representations

$A_2 + B_1 + B_2$

Much of the use of Group Theory to solve real problems involves generating a reducible representation, and then reducing it to its constituent irreducible representations.

In the example above this could be done by inspection, but many examples are far too complex, and a REDUCTION FORMULA has to be used. This formula is:

Number of times an
irreducible representation $= \dfrac{1}{h} \displaystyle\sum_{\substack{\text{over all} \\ \text{classes}}} \chi_R \times \chi_I \times N$
occurs in the reducible
representation

where h = order of the group (= number of operations in
the group)

χ_R = character of the reducible representation

χ_I = character of the irreducible representation

N = number of symmetry operations in the class
(i.e. the number of equivalent operations. See
frames 2.35–2.40)

In the example in frame 3.17 h = 4, χ_R = 3 for E, 3 for C_2,
and 1 for each σ.

For the A_1 representation χ_I is 1 for each operation, hence:

$$\text{Number of } A_1 = \tfrac{1}{4} [\underbrace{3 \times 1 \times 1}_{E} + \underbrace{3 \times 1 \times 1}_{C_2}$$

$$+ \underbrace{1 \times 1 \times 1}_{\sigma(xz)} + \underbrace{1 \times 1 \times 1}_{\sigma(yz)}] = 2$$

For the A_2 representation, the values of χ_I are 1, 1, −1, −1,
hence:

$$\text{Number of } A_2 = \tfrac{1}{4} [\underbrace{(3 \times 1 \times 1)}_{E} + \underbrace{(3 \times 1 \times 1)}_{C_2}$$

$$+ \underbrace{(1 \times (-1) \times 1)}_{\sigma(xz)} + \underbrace{(1 \times (-1) \times 1)}_{\sigma(yz)}] = 1$$

Do the same thing to find the number of B_1 and B_2 species.

3.19   Number of $B_1 = \frac{1}{4}[(3 \times 1 \times 1) + (3 \times (-1) \times 1)$
$$+ (1 \times 1 \times 1) + (1 \times (-1) \times 1)] = 0$$

Number of $B_2 = \frac{1}{4}[(3 \times 1 \times 1) + (3 \times (-1) \times 1)$
$$+ (1 \times (-1) \times 1) + (1 \times 1 \times 1)] = 0$$

i.e the reducible representation reduces to $2A_1 + A_2$.

Let us consider the representation of C_{3v} labelled Γ_1, (reducible representations are commonly designated by a capital gamma, Γ):

C_{3v}	E	$2C_3$	$3\sigma_v$
Γ_1	4	1	-2

In this case, the number of operations in the class ($=$ N in the formula) is two for the rotations and three for the reflections. The reduction is therefore performed using the character table as follows:

	E	$2C_3$	$3\sigma_v$
Γ_1	4	①	-2

N.B. Do not worry about the figure 2 in the character table — its significance will be come clear later. I apologise for the nomenclature which uses E for a representation and for the identity but it is a standard convention.

C_{3v}	E	②C_3	$3\sigma_v$
A_1	1	①	1
A_2	1	1	-1
E	2	-1	0

Number of $A_1 = \frac{1}{6}[(4 \times 1 \times 1) + (1 \times 1 \times 2) + (-2 \times 1 \times 3)] = 0$

Number of $A_2 = \frac{1}{6}[(4 \times 1 \times 1) + (1 \times 1 \times 2) + (-2 \times -1 \times 3)] = 2$

Number of E $= ?$

3.20 $\frac{1}{6}[(4 \times 2 \times 1) + (1 \times -1 \times 2) + (-2 \times 0 \times 3)] = 1$

i.e. Γ_1 reduces to $2A_2 + E$

Confirm this by adding these representations.

3.21

A_2	1	1	-1
A_2	1	1	-1
E	2	-1	0
$2A_2 + E$	4	1	-2

The next few frames are practice at the very vital business of reducing reducible representations. For this you should use the character tables printed at the back of the book.

Reduce the representation (C_{3v})

	E	$2C_3$	$3\sigma_v$
Γ_2	4	1	0

3.22   Number of $A_1 = \frac{1}{6}[(4 \times 1 \times 1) + (1 \times 1 \times 2) + 0] = 1$

$A_2 = \frac{1}{6}[(4 \times 1 \times 1) + (1 \times 1 \times 2) + 0] = 1$

$E = \frac{1}{6}[(4 \times 2 \times 1) + (1 \times -1 \times 2) + 0] = 1$

$\Gamma_2 = A_1 + A_2 + E$

What is the order h of the C_{2v} and C_{2h} groups?

3.23   4 in each case, i.e. both groups have 4 operations.
Reduce the representation:

C_{2v}	E	C_2	$\sigma(xz)$	$\sigma(yz)$
Γ_3	2	0	0	-2

3.24   Number of $A_1 = \frac{1}{4}[(2 \times 1 \times 1) + 0 + 0 + (-2 \times 1 \times 1)] = 0$

$A_2 = \frac{1}{4}[(2 \times 1 \times 1) + 0 + 0 + (-2 \times -1 \times 1)] = 1$

$B_1 = \frac{1}{4}[(2 \times 1 \times 1) + 0 + 0 + (-2 \times -1 \times 1)] = 1$

$B_2 = \frac{1}{4}[(2 \times 1 \times 1) + 0 + 0 + (-2 \times 1 \times 1)] = 0$

$$\Gamma_3 = A_2 + B_1$$

As mentioned earlier, the reduction of reducible representations is vital to the use of group theory. The following six examples are included for practice and can be omitted if you feel really confident.

Reduce the following reducible representations:

C_{2v}	E	C_3	$\sigma(xz)$	$\sigma(yz)$
Γ_4	3	1	-1	-3
Γ_5	30	0	0	10

C_{2h}	E	C_2	i	σ_h
Γ_6	2	0	-2	0
Γ_7	30	0	0	10

C_{3v}	E	$2C_3$	$3\sigma_v$
Γ_8	5	2	1
Γ_9	7	-2	1

3.25 $\Gamma_4 = 2A_2 + B_1$

$\Gamma_5 = 10A_1 + 5A_2 + 5B_1 + 10B_2$

$\Gamma_6 = A_u + B_u$

$\Gamma_7 = 10A_g + 5B_g + 5A_u + 10B_u$

$\Gamma_8 = 2A_1 + A_2 + E$

$\Gamma_9 = A_1 + 3E$

Let us now turn to the group C_{4v} of which the following complex is an example:

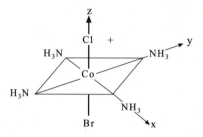

What are the operations of the C_{4v} group? (Remember that an axis can generate several operations.)

3.26   E C_4 C_4^2 C_4^3

Two vertical planes passing through the NH_3 groups (σ_v) and two vertical planes passing between the NH_3 groups (σ_d).

We usually group the operations in classes as:

E $2C_4$ $C_2(= C_4^2)$ $2\sigma_v$ $2\sigma_d$

Taking the z axis as being vertical, what number represents the operation C_4 on an arrow in the z direction?

3.27   1. i.e. z is symmetric to C_4
        What numbers represent the effect of the other operations on z?

3.28 | E | $2C_4$ | C_2 | $2\sigma_v$ | $2\sigma_d$ |
        |---|--------|-------|-------------|-------------|
        | 1 | 1 | 1 | 1 | 1 |

i.e. z belongs to the totally symmetric or A_1 representation of the C_{4v} group.

What happens to an arrow along the y axis when a C_4 operation is performed on it clockwise?

3.29   It points along the x axis, i.e. y is converted to x by C_4. Now we have problems! There is no simple number which will convert y to x (and also x to $-y$), so the representation cannot be a simple number. The only way to represent the transformations $x \rightarrow -y$ and $y \rightarrow x$ is to use a matrix, and the next programme is about matrices as representations of operations.

We can, however, draw a useful conclusion at this stage from a simple symmetry argument. What effect does application of the C_4 operation have on the total energy of the $[Co(NH_3)_4ClBr]^+$ ion?

3.30   None at all. If C_4 is a symmetry operation, it leaves the molecule indistinguishable, and that includes its energy.

What happens to the p_y orbital on application of a clockwise C_4 about the z axis?

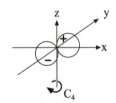

3.31   It becomes a p_x orbital:

If application of a symmetry operation does not change the total energy but interconverts two orbitals, what can we say about the energies of the two orbitals?

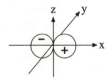

3.32   They must be identical, i.e. degenerate.

We will be seeing that the p_x and p_y orbitals both belong to the same DEGENERATE REPRESENTATION of C_{4v}, and this indicates directly that the two orbitals are degenerate. So far we have only been looking at non-degenerate representations — hence the title of the programme.

Are the p_x and p_y orbitals degenerate in C_{2v} symmetry? Look at a C_{2v} character table to see the representations to which x and y belong.

3.33   The two orbitals are not degenerate in C_{2v} because x belongs to B_1 and y to B_2.

In this case they belong to different representations, and we can tell from symmetry alone that p_x and p_y are of different energy in a C_{2v} molecule. This can be seen readily for the water molecule because one orbital is largely in the molecular plane, and the other is out of it. Their energies will therefore be affected to a different extent by the two hydrogen atoms:

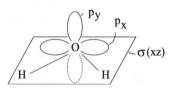

Symmetry alone will never tell us the extent of any energy split, it will only tell us if the energy difference is precisely zero (p_x and p_y in C_{4v}) or not zero (p_x and p_y in C_{2v}). In the same way we can use symmetry to find if a spectroscopic transition has a finite probability (is allowed) or has a precisely zero probability (is forbidden). Symmetry will not tell us the intensity of the transition, i.e. it will not tell us the actual value of the probability, only that it is or is not zero.

You should now be able to form a simple non degenerate representation to describe the effect of the symmetry operations of a group on a direction such as x, and you should be able to reduce a reducible representation to its component irreducible representations. The importance of being able to reduce a reducible representation cannot be over emphasised.

There now follows a short test to show you how well you can form simple representations and reduce less simple ones.

Non-degenerate Representations Test

The C_{2h} character table is, in part:

C_{2h}	E	C_2	i	σ_h
A_g	1	1	1	1
B_g	1	−1	1	−1
A_u	1	1	−1	−1
B_u	1	−1	−1	1

1.a   Taking the C_2 axis as the z axis, and σ_h to be the xy plane, to what representations do x, y, and z belong in C_{2h} symmetry?

1.b   To what representations do the d_{xy}, d_{xz} and d_{yz} orbitals belong in C_{2h} symmetry?

2.   Reduce the following reducible representations:

C_{2h}	E	C_2	i	σ_h
Γ_{10}	8	0	6	2
Γ_{11}	3	1	−3	−1

C_{3v}	E	$2C_3$	$3\sigma_v$
Γ_{12}	6	0	−2
Γ_{13}	9	0	−1

C_{2v}	E	C_2	$\sigma(xz)$	$\sigma(yz)$
Γ_{14}	3	−3	1	−1
Γ_{15}	17	3	−13	1

Answers

1.a   x belongs to B_u *1 mark*

        y belongs to B_u *1 mark*

        z belongs to A_u *1 mark*

  b   xy belongs to A_g *1 mark*

        xz belongs to B_g *1 mark*

        yz belongs to B_g *1 mark*

2. $\Gamma_{10} = 4A_g + 3B_g + B_u$ *1 mark*

       $\Gamma_{11} = 2A_u + B_u$ *1 mark*

       $\Gamma_{12} = 2A_2 + 2E$ *1 mark*

       $\Gamma_{13} = A_1 + 2A_2 + 3E$ *1 mark*

       $\Gamma_{14} = 2B_1 + B_2$ *1 mark*

       $\Gamma_{15} = 2A_1 + 8A_2 + 7B_2$ *1 mark*

                             Total   *12 marks*

Before you proceed to the next programme you should have obtained at least:

Question 1   (objective 1)   3/6   (Frames 3.2–3.16)
Question 2   (objective 2)   5/6   (Frames 3.17–3.24)

If you have not obtained this score on question 2 in particular, you would be well advised to return to the frames shown. Ask somebody to construct some reducible representations for you (by adding irreducible representations), and practice the use of the reduction formula until you have mastered it.

Non-degenerate Representations

Revision Notes

The symmetry operations of a group can be represented by sets of numbers termed irreducible representations which:

i.   represent the effect of the group operations on certain directional properties e.g. x xz R_x etc.

ii.   multiply together in the same way as the group operations.

The use of group theory frequently involves producing a reducible representation which is the sum of a number of the irreducible representations in the character table. This reducible representation then has to be reduced to its component irreducible representations either by inspection or by using the reduction formula:

Number of times an
irreducible representation
occurs in the reducible
representation

$$= \frac{1}{h} \sum_{\substack{\text{over all} \\ \text{classes}}} \chi_R \times \chi_I \times N$$

where h = the order of the group (= number of operations in the group)

χ_R = character of the reducible representation

χ_I = character of the irreducible representation

N = number of symmetry operations in the class

In some point groups (those with proper axes of order greater than 2), a symmetry operation causes two directional properties to mix. These directional properties must then be degenerate, and the operation must be represented by a matrix, termed a degenerate representation.

Matrices

Objectives

After completing this programme you should be able to:

1. Combine two matrices.
2. Set up a matrix to perform a given transformation.
3. Find the character of a matrix representing a symmetry operation, using any given basis.

All three objectives are tested at the end of the programme.

Assumed Knowledge

You should be able to plot a point, or visualise how it is plotted, in three dimensions, i.e. given x, y and z co-ordinates.

Matrices

4.1   We left the previous programme on representations at the
point where a symmetry operation had the effect of inter-
converting x and y. Such an operation cannot be represented
by a single number, but we shall see in this programme that
the operation can easily be represented by a matrix. The
programme will not go deeply into the subject of matrix
algebra but it will be necessary to learn how to combine
two matrices so that the effect of two successive symmetry
operations can be represented in matrix form.

A matrix is an array of numbers enclosed within either
square or rounded brackets, e.g.

$$\begin{bmatrix} 1 & 4 & 7 \\ 2 & -6 & 3 \\ 8 & 0 & 5 \end{bmatrix} \quad \text{or} \quad \begin{pmatrix} 1 & 0 \\ 0 & -1 \end{pmatrix}$$

Each number is termed an *element* of the matrix.

These are examples of *square matrices* because the number of
columns equals the number of *rows* in each case, but a matrix
may have any number of columns or rows.

A matrix, unlike a determinant, does not have a numerical
value – its use is in the effect it has on another matrix which
can represent a point or a direction.

Write down a *one column matrix* to represent the co-
ordinates of the point (3, 1, 2) i.e. $x = 3$ $y = 1$ $z = 2$.

4.2 $$\begin{pmatrix} 3 \\ 1 \\ 2 \end{pmatrix} \quad \text{or} \quad \begin{bmatrix} 3 \\ 1 \\ 2 \end{bmatrix}$$

This column matrix represents either the co-ordinates (3, 1, 2)
or a line (vector) starting at the origin and finishing at (3, 1, 2).
We shall be looking at the effect of rotating this line about
the z axis, and the way in which matrices can represent the
rotations.

Write down a *row matrix* representing the vector from the
origin to (3, 1, 2).

4.3   (3   1   2)

Note that the matrix has no commas, unlike the set of co-ordinates.

If we can convert our matrix $\begin{pmatrix} 3 \\ 1 \\ 2 \end{pmatrix}$ to the matrix $\begin{pmatrix} -3 \\ -1 \\ 2 \end{pmatrix}$ we

shall have changed our line to one pointing from the origin to the point (−3, −1   2). Looking down the z axis, our original column matrix represents the line OA:

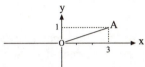

Draw the line OA′ represented by the new matrix $\begin{pmatrix} -3 \\ -1 \\ 2 \end{pmatrix}$

4.4

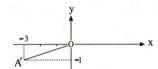

The line OA′ can be obtained from OA by rotating OA by half a turn about the z axis. Thus whatever it is that changes

the matrix $\begin{pmatrix} 3 \\ 1 \\ 2 \end{pmatrix}$ to $\begin{pmatrix} -3 \\ -1 \\ 2 \end{pmatrix}$ can be said to *represent* the

operation of rotation by half a turn about the z axis.

Draw the line OA″ obtained by rotating OA by $\frac{1}{4}$ turn (clockwise) about the z axis.

4.5

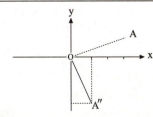

What is the value of the new x co-ordinate?

4.6   1 i.e. the new x co-ordinate is the same as the old y co-ordinate.
        What is the value of the new y co-ordinate?

___

4.7   −3 i.e. the new y co-ordinate is minus the old x co-ordinate.
        What, then, is the matrix representing OA″?

___

4.8 $$\begin{pmatrix} 1 \\ -3 \\ 2 \end{pmatrix}$$

        We can make this more general by saying that the new x co-
        ordinate equals the original y, the new y co-ordinate equals
        minus the original x and the z co-ordinate is left alone. The
        new co-ordinates are therefore (y, −x, z).

        Write down the matrix representing the general set of new
        co-ordinates.

___

4.9 $$\begin{pmatrix} y \\ -x \\ z \end{pmatrix}$$

        Thus in the general case, the operation of a $\frac{1}{4}$ turn rotation

        can be represented by a matrix M where $M \begin{pmatrix} x \\ y \\ z \end{pmatrix} = \begin{pmatrix} y \\ -x \\ z \end{pmatrix}$

        The matrix M is then a representation of the C_4 rotation in
        the same way as we used +1 and −1 as representations in the
        previous programme.

        The equation above raises two questions which will now be
        examined:

        a. How can matrices be combined?
        b. How can a matrix like M be set up?

        Matrices can be combined or multiplied provided the two
        matrices are *conformable*. Two matrices (x) and (y) are con-
        formable if the number of columns in (x) is equal to the
        number of rows in (y).

        Write down a suitable matrix (y) if matrix (x) is $\begin{pmatrix} a & b & c \\ d & e & f \end{pmatrix}$

___

4.10
$$\begin{pmatrix} g & h & \dots \\ i & j & \dots \\ k & l & \dots \end{pmatrix}$$ or any other 3-row matrix.

The product of any two matrices is easily formed by remembering the letters R C. An element in the rth row and the cth column of the product is formed by multiplying together the elements from the rth row of matrix 1 and the cth column of matrix 2 and summing the products, e.g.

$$\begin{pmatrix} a & b & c \\ d & e & f \end{pmatrix} \begin{pmatrix} r & s & t \\ u & v & w \\ x & y & z \end{pmatrix} = \begin{pmatrix} A & B & C \\ D & E & F \end{pmatrix}$$

Note that the product matrix has two rows (the same as the first matrix) and three columns (the same as the second matrix). This result is quite general.

The value of the element A which is in Row 1 and Column 1 of the product is obtained by working along Row 1 of the first matrix, down Column 1 of the second; and summing the products.

Row 1 of 1st matrix

$A = (a \times r) + (b \times u) + (c \times x)$

Column 1 of 2nd matrix.

What is the value of element D in Row 2, Column 1 of the product?

4.11   Row 2 of 1st matrix

$D = (d \times r) + (e \times u) + (f \times x)$

Column 1 of 2nd matrix

What is the value of element E?

4.12 $E = (d \times s) + (e \times v) + (f \times y)$

You should now be able to write down the whole of the product matrix.

4.13

$$\text{Product} = \begin{pmatrix} ar + bu + cx & as + bv + cy & at + bw + cz \\ dr + eu + fx & ds + ev + fy & dt + ew + fz \end{pmatrix}$$

Now a simple numerical example:

$$\begin{pmatrix} 1 & 2 \\ 3 & 4 \end{pmatrix} \quad \begin{pmatrix} 5 & 6 & 7 \\ 8 & 9 & 10 \end{pmatrix} =$$

Row 1 of 1st matrix

$$(1 \times 5 + 2 \times 8 \quad 1 \times 6 + 2 \times 9 \quad 1 \times 7 + 2 \times 10)$$

Column 3 of 2nd matrix

Complete the second row of this matrix.

4.14

$$\begin{pmatrix} 21 & 24 & 27 \\ 3 \times 5 + 4 \times 8 & 3 \times 6 + 4 \times 9 & 3 \times 7 + 4 \times 10 \end{pmatrix}$$

$$= \begin{pmatrix} 21 & 24 & 27 \\ 47 & 54 & 61 \end{pmatrix} \quad \text{Row 2 of 1st} \quad \text{Column 3 of 2nd}$$

Calculate the product: $\begin{pmatrix} 1 & 2 \\ 3 & 4 \end{pmatrix} \begin{pmatrix} 1 & 1 \\ 2 & 2 \end{pmatrix}$

4.15

$$\begin{pmatrix} 5 & 5 \\ 11 & 11 \end{pmatrix}$$

Now try them the other way round:

$$\begin{pmatrix} 1 & 1 \\ 2 & 2 \end{pmatrix} \quad \begin{pmatrix} 1 & 2 \\ 3 & 4 \end{pmatrix} =$$

4.16

$$\begin{pmatrix} 4 & 6 \\ 8 & 12 \end{pmatrix}$$ i.e. the order of multiplication affects the result.

This is quite common. If the order of multiplication is important, the matrices are said not to *commute*. In some cases the order of the matrices does not affect the result, in which case they do commute or are *commutative matrices*.

One clear case of non commutation occurs with the matrices $\begin{pmatrix} 1 \\ 2 \\ 3 \end{pmatrix}$ and $(3 \quad 2 \quad 1)$

Remember that the product has the same number of rows as the first matrix and the same number of columns as the second. How many rows and columns are there in the product:

$$\begin{pmatrix} 1 \\ 2 \\ 3 \end{pmatrix} \quad (3 \quad 2 \quad 1)?$$

4.17   3 rows (same as 1st matrix).
       3 columns (same as 2nd matrix).

When evaluating this product, there is only one element in each row of matrix 1 and only one element in each column of matrix 2, so no addition is necessary.

Evaluate $\begin{pmatrix} 1 \\ 2 \\ 3 \end{pmatrix} \quad (3 \quad 2 \quad 1)$

4.18 $\begin{pmatrix} 3 & 2 & 1 \\ 6 & 4 & 2 \\ 9 & 6 & 3 \end{pmatrix}$

Now try them the other way round: $(3 \quad 2 \quad 1) \quad \begin{pmatrix} 1 \\ 2 \\ 3 \end{pmatrix}$

How may rows and columns will the product have?

4.19   1 row and 1 column, i.e. it will be a single number.

Evaluate $(3 \quad 2 \quad 1) \quad \begin{pmatrix} 1 \\ 2 \\ 3 \end{pmatrix}$

4.20 $(3 \times 1 + 2 \times 2 + 1 \times 3) = (10)$

Evaluate the product:

$$\begin{pmatrix} 0 & 1 & 0 \\ -1 & 0 & 0 \\ 0 & 0 & 1 \end{pmatrix} \quad \begin{pmatrix} 3 \\ 1 \\ 2 \end{pmatrix}$$

4.21 $\begin{pmatrix} 1 \\ -3 \\ 2 \end{pmatrix}$

i.e. the matrix $\begin{pmatrix} 0 & 1 & 0 \\ -1 & 0 & 0 \\ 0 & 0 & 1 \end{pmatrix}$ represents one of the

operations on the line OA. Which one? (see frame 4.5)

4.22 Clockwise rotation by $\frac{1}{4}$ turn about z i.e. OA becomes OA″:

Evaluate the product:

$\begin{pmatrix} 0 & 1 & 0 \\ -1 & 0 & 0 \\ 0 & 0 & 1 \end{pmatrix} \begin{pmatrix} x \\ y \\ z \end{pmatrix}$

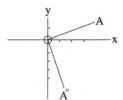

4.23 $\begin{pmatrix} y \\ -x \\ z \end{pmatrix}$

i.e. our matrix converts x to y, and y to −x in any general case. It is therefore a quite general representation of the $\frac{1}{4}$ turn operation, and is not specific to the set of co-ordinates (3, 1, 2).

Evaluate the products:

$\begin{pmatrix} -1 & 0 & 0 \\ 0 & -1 & 0 \\ 0 & 0 & 1 \end{pmatrix} \begin{pmatrix} 3 \\ 1 \\ 2 \end{pmatrix}$ and $\begin{pmatrix} -1 & 0 & 0 \\ 0 & -1 & 0 \\ 0 & 0 & 1 \end{pmatrix} \begin{pmatrix} x \\ y \\ z \end{pmatrix}$

4.24 $\begin{pmatrix} -3 \\ -1 \\ 2 \end{pmatrix}$ and $\begin{pmatrix} -x \\ -y \\ z \end{pmatrix}$

i.e. the matrix $\begin{pmatrix} -1 & 0 & 0 \\ 0 & -1 & 0 \\ 0 & 0 & 1 \end{pmatrix}$ is a general representation

of an operation on OA. Which operation?

4.25   Rotation by half a turn about z.

We will now turn to the second question raised in frame 4.9, namely how can we generate a matrix which will perform the

required operation on $\begin{pmatrix} x \\ y \\ z \end{pmatrix}$? This is very simple if we write

in symbolic form the statements:

"New x becomes $-1 \times$ old x + zero times old y + zero $\times$ old z"

or: $x = -1x + 0y + 0z$ etc.

For the $\frac{1}{2}$ turn operation, the full set of equations is:

$$x = (-1)x + \quad 0y + 0z$$
$$y = \quad 0x + (-1)y + 0z$$
$$z = \quad 0x + \quad 0y + 1z$$

And the matrix can be written down by inspection as

$$\begin{pmatrix} -1 & 0 & 0 \\ 0 & -1 & 0 \\ 0 & 0 & 1 \end{pmatrix}$$

For a clockwise rotation of $\frac{1}{4}$ of a turn about the z axis, the new x co-ordinate is the same as the old y co-ordinate. Work out the values of the new y and z co-ordinates and write out the equations for the rotation.

4.26 $$x = \quad 0x + 1y + 0z$$
$$y = -1x + 0y + 0z$$
$$z = \quad 0x + 0y + 1z$$

Work out the effect on the x, y and z co-ordinates of reflection in the xy plane, and hence write out the set of equations for this reflection operation.

4.27 $$x = \quad x + 0y + \quad 0z$$
$$y = \quad 0x + 1y + \quad 0z$$
$$z = \quad 0x + 0y + -1z$$

Because the reflection changes the sign of z, but leaves x and y unchanged.

What is the corresponding matrix?

4.28
$$\begin{pmatrix} 1 & 0 & 0 \\ 0 & 1 & 0 \\ 0 & 0 & -1 \end{pmatrix}$$

Write out the full matrix equation showing the operation of reflection in the xy plane on the point (x, y, z).

4.29
$$\begin{pmatrix} 1 & 0 & 0 \\ 0 & 1 & 0 \\ 0 & 0 & -1 \end{pmatrix} \begin{pmatrix} x \\ y \\ z \end{pmatrix} = \begin{pmatrix} x \\ y \\ -z \end{pmatrix}$$

Matrix algebra is a fairly complex subject but it is not necessary to go into it in great detail for our present purposes. We shall, however, be making use of some of the results which come from a study of matrix algebra and many of these results can be expressed in terms of the character of a square matrix. The character (sometimes called the trace) of a square matrix is simply the sum of the diagonal elements from top left to bottom right.

What is the character of:

$$\begin{pmatrix} 1 & 0 & 0 \\ 0 & 1 & 0 \\ 0 & 0 & -1 \end{pmatrix} \text{ and } \begin{pmatrix} 1 & 0 \\ 0 & -1 \end{pmatrix} \text{ and } \begin{pmatrix} -1 & 0 \\ -2 & -1 \end{pmatrix} \text{ and } \begin{pmatrix} -1 & 0 \\ 0 & -1 \end{pmatrix}$$

4.30   1   0   −2   −2

Express the following transformation in matrix form, and work out the character of the matrix:

$$x = \frac{\sqrt{3}x}{2} + \frac{-1y}{2}$$

$$y = \frac{1x}{2} + \frac{\sqrt{3}y}{2}$$

4.31 $\sqrt{3}$

i.e. the matrix is $\begin{pmatrix} \dfrac{\sqrt{3}}{2} & -\dfrac{1}{2} \\ \dfrac{1}{2} & \dfrac{\sqrt{3}}{2} \end{pmatrix}$ and the character is

$$\frac{\sqrt{3}}{2} + \frac{\sqrt{3}}{2} = \sqrt{3}$$

It should be clear that the character is dependent only on the two terms $\sqrt{3}/2$ which express the extent to which x is converted to itself and y is converted to itself in the original two equations. This result is very important and will allow us to greatly simplify much of the routine application of group theory.

Use this result to write down the character of the matrix representing the transformation:

a = 2a + ... + 10d
b = ... + 6b + ...
c = ... − 4c + ...
d = ... + 3d

4.32   Character = 7

          = 2 + 6 − 4 + 3   i.e. it depends only on the extent
                                    to which a is converted to a,
                                    b to b, etc.

We have so far used cartesian co-ordinates to generate matrices representing operations, but we can use other terms e.g. we can represent the operation of a half turn rotation on the O–H bonds of water as:

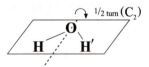

H′ becomes H i.e. new H′ = 0 × old H′ + 1 × old H
H becomes H′ etc.

$$M \begin{pmatrix} H' \\ H \end{pmatrix} = \begin{pmatrix} H \\ H' \end{pmatrix}$$

What is the matrix M representing the transformation?

4.33 $\begin{pmatrix} 0 & 1 \\ 1 & 0 \end{pmatrix}$

i.e. $\begin{pmatrix} 0 & 1 \\ 1 & 0 \end{pmatrix} \begin{pmatrix} H' \\ H \end{pmatrix} = \begin{pmatrix} H \\ H' \end{pmatrix}$

We say that the O–H bonds have been used as a basis for a representation of the rotation.

Use the small arrows shown as a basis for the same half turn rotation.

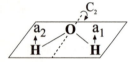

Hint:
The positive direction
of the arrows is upwards.

4.34 $\begin{pmatrix} 0 & -1 \\ -1 & 0 \end{pmatrix}$

i.e. new $a_1 = -$old a_2 (pointing the other way)
     new $a_2 = -$old a_1

$\begin{pmatrix} 0 & -1 \\ -1 & 0 \end{pmatrix} \begin{pmatrix} a_1 \\ a_2 \end{pmatrix} = \begin{pmatrix} -a_2 \\ -a_1 \end{pmatrix}$

Use a_1 and a_2 as a basis for a representation of a reflection in the molecular plane.

4.35 $\begin{pmatrix} -1 & 0 \\ 0 & -1 \end{pmatrix}$

i.e. $\begin{pmatrix} -1 & 0 \\ 0 & -1 \end{pmatrix} \begin{pmatrix} a_1 \\ a_2 \end{pmatrix} = \begin{pmatrix} -a_1 \\ -a_2 \end{pmatrix}$

What is the character of this representation of reflection?

4.36   Character $= -2$

When considering molecular vibrations it is necessary to
work out the *cartesian representation* by using the x, y and
z directions on each atom as a basis. This basis for the water
molecule, looks like:

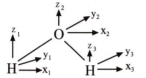

The molecular plane
is the xz plane.

If we apply a $\frac{1}{2}$ turn rotation about z_2, then the new x_1 equals
$-x_3$, the new y_1 equals $-y_3$, the new z_1 equals z_3 etc.

The half turn rotation will be represented by a 9×9 matrix
which carries out all these transformations i.e.

$$M \begin{pmatrix} x_1 \\ y_1 \\ z_1 \\ x_2 \\ y_2 \\ z_2 \\ x_3 \\ y_3 \\ z_3 \end{pmatrix} = \begin{pmatrix} -x_3 \\ -y_3 \\ z_3 \\ \\ \text{etc.} \\ \\ \\ \\ \end{pmatrix}$$

What is the character of the 9×9 matrix M? If you can work
this out by using the important simplification in frame 4.31
then do so. The answer gives the full matrix equation for the
transformation.

4.37   Character $= -1$.

The arrows on hydrogen are completely moved, and contribute nothing to the character.

x_2 and y_2 are reversed and contribute -1 each.

z_2 is unaffected and contributes $+1$.

The full equation is:

$$\begin{pmatrix} 0 & 0 & 0 & 0 & 0 & 0 & -1 & 0 & 0 \\ 0 & 0 & 0 & 0 & 0 & 0 & 0 & -1 & 0 \\ 0 & 0 & 0 & 0 & 0 & 0 & 0 & 0 & 1 \\ 0 & 0 & 0 & -1 & 0 & 0 & 0 & 0 & 0 \\ 0 & 0 & 0 & 0 & -1 & 0 & 0 & 0 & 0 \\ 0 & 0 & 0 & 0 & 0 & 1 & 0 & 0 & 0 \\ -1 & 0 & 0 & 0 & 0 & 0 & 0 & 0 & 0 \\ 0 & -1 & 0 & 0 & 0 & 0 & 0 & 0 & 0 \\ 0 & 0 & 1 & 0 & 0 & 0 & 0 & 0 & 0 \end{pmatrix} \begin{pmatrix} x_1 \\ y_1 \\ z_1 \\ x_2 \\ y_2 \\ z_2 \\ x_3 \\ y_3 \\ z_3 \end{pmatrix} = \begin{pmatrix} -x_3 \\ -y_3 \\ +z_3 \\ -x_2 \\ -y_2 \\ +z_2 \\ -x_1 \\ -y_1 \\ +z_1 \end{pmatrix}$$

It is clearly an advantage not to have to write out the whole matrix if at all possible!

The number of possible representations of an operation is clearly very large, and depends only on our ingenuity in devising bases to generate representations.

Generate a representation of the two fold rotation, using the four arrows shown as the basis:

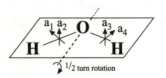

The full matrix equation is shown in the answer.
(N.B. a_2 and a_3 are perpendicular to the plane, a_1 and a_4 are in it.)

4.38   Character = 0 (all four arrows are shifted by the operation)

$$\begin{pmatrix} 0 & 0 & 0 & 1 \\ 0 & 0 & -1 & 0 \\ 0 & -1 & 0 & 0 \\ 1 & 0 & 0 & 0 \end{pmatrix} \begin{pmatrix} a_1 \\ a_2 \\ a_3 \\ a_4 \end{pmatrix} = \begin{pmatrix} a_4 \\ -a_3 \\ -a_2 \\ a_1 \end{pmatrix} \quad \text{Character} = 0$$

Use the same four arrows as the basis of a representation of the operation of reflection in the plane of the molecule. Write out the matrix equation and find the character of the representation.

4.39 $$\begin{pmatrix} 1 & 0 & 0 & 0 \\ 0 & -1 & 0 & 0 \\ 0 & 0 & -1 & 0 \\ 0 & 0 & 0 & 1 \end{pmatrix} \begin{pmatrix} a_1 \\ a_2 \\ a_3 \\ a_4 \end{pmatrix} = \begin{pmatrix} a_1 \\ -a_2 \\ -a_3 \\ a_4 \end{pmatrix} \quad \text{Character} = 0$$

You should now be able to:

i.   Combine two matrices.
ii.   Set up a matrix to perform a certain transformation.
iii. Find the character of a matrix representing an operation, using any given basis.

All of these are important in the application of molecular symmetry to a wide range of problems.

The following test should show you how much you have learned about matrices.

Matrices Test

1. What is meant by the statement "Two matrices (A) and (B) commute"?

2. Show whether or not the matrices $\begin{pmatrix} 1 & 2 \\ 0 & 1 \end{pmatrix}$ and $\begin{pmatrix} 0 & 2 \\ 2 & 0 \end{pmatrix}$ commute

3. Which of the following can be combined with $\begin{pmatrix} a & b \\ c & d \\ e & f \end{pmatrix}$:

   A) $\begin{pmatrix} 1 & 2 & 3 & 4 \\ 5 & 6 & 7 & 8 \end{pmatrix}$ D) $\begin{pmatrix} 1 & 4 \\ 2 & 5 \\ 3 & 6 \end{pmatrix}$

   B) $\begin{pmatrix} 1 & 5 \\ 2 & 6 \\ 3 & 7 \\ 4 & 8 \end{pmatrix}$ E) $\begin{pmatrix} 1 & 2 \\ 3 & 4 \end{pmatrix}$

   C) $\begin{pmatrix} 1 & 2 & 3 \\ 4 & 5 & 6 \end{pmatrix}$

4. Combine the following matrices:

   A) $\begin{pmatrix} 1 & 0 & 2 \\ 4 & 1 & -3 \\ 2 & 3 & 0 \end{pmatrix} \begin{pmatrix} 2 & 1 & 4 \\ 3 & 0 & 6 \\ 0 & -1 & 2 \end{pmatrix} =$

   B) $\begin{pmatrix} 2 & 0 & 0 \\ 1 & 0 & 0 \\ 0 & 0 & 4 \end{pmatrix} \begin{pmatrix} x \\ y \\ z \end{pmatrix} =$

   C) $\begin{pmatrix} 1 & 6 \\ 2 & 8 \end{pmatrix} \begin{pmatrix} 4 & 0 \\ 3 & 7 \end{pmatrix} =$

5.   Set up the matrices which will perform the following transformations:

A) $\begin{pmatrix} x \\ y \end{pmatrix}$ to $\begin{pmatrix} -y \\ -x \end{pmatrix}$

B) $\begin{pmatrix} x \\ y \end{pmatrix}$ to $\begin{pmatrix} x \\ y \end{pmatrix}$ (i.e. leave the original unchanged)

C) $\begin{pmatrix} x \\ y \\ z \end{pmatrix}$ to $\begin{pmatrix} -\sqrt{2}y \\ \sqrt{2}y \\ -x \end{pmatrix}$

D) $\begin{pmatrix} x \\ y \\ z \end{pmatrix}$ to $\begin{pmatrix} -y \\ x \\ -z \end{pmatrix}$

6.   Write down the character of each of the matrices derived in question 5.

Use the following diagrams for questions 7, 8 and 9:

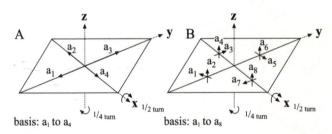

basis: a_1 to a_4 basis: a_1 to a_8

7.   Write down the characters of the matrices representing the quarter turn rotation, using the bases A and B shown.

8.   Write down the characters of the matrices representing the operation of reflection in the xy plane, using the bases shown.

9.   Write down the characters of the matrices representing the half turn rotation about the x axis using the bases shown.

Answers

1. $(A)(B) = (B)(A)$ *1 mark*

2. $\begin{pmatrix} 1 & 2 \\ 0 & 1 \end{pmatrix} \begin{pmatrix} 0 & 2 \\ 2 & 0 \end{pmatrix} = \begin{pmatrix} 4 & 2 \\ 2 & 0 \end{pmatrix}$ *2 marks*

$\begin{pmatrix} 0 & 2 \\ 2 & 0 \end{pmatrix} \begin{pmatrix} 1 & 2 \\ 0 & 1 \end{pmatrix} = \begin{pmatrix} 0 & 2 \\ 2 & 4 \end{pmatrix}$ They do not commute

3.   Any matrix with two rows, i.e. A,   *1 mark*
                                        C,   *1 mark*
                                        E.   *1 mark*

4.   A) $\begin{pmatrix} 2 & -1 & 8 \\ 11 & 7 & 16 \\ 13 & 2 & 26 \end{pmatrix}$ *1 mark*

B) $\begin{pmatrix} 2x \\ x \\ 4z \end{pmatrix}$ *1 mark*

C) $\begin{pmatrix} 22 & 42 \\ 32 & 56 \end{pmatrix}$ *1 mark*

5.   A) $\begin{pmatrix} 0 & -1 \\ -1 & 0 \end{pmatrix}$ *1 mark*

B) $\begin{pmatrix} 1 & 0 \\ 0 & 1 \end{pmatrix}$ *1 mark*

C) $\begin{pmatrix} 0 & -\sqrt{2} & 0 \\ 0 & \sqrt{2} & 0 \\ -1 & 0 & 0 \end{pmatrix}$ *1 mark*

D) $\begin{pmatrix} 0 & -1 & 0 \\ 1 & 0 & 0 \\ 0 & 0 & -1 \end{pmatrix}$ *1 mark*

6.   A   0   *1 mark*
        B   2
        C $\sqrt{2}$
        D -1

7.   A   0   *1 mark*
        B   0   *1 mark*

8.   A   4   *1 mark*
        B   0   *1 mark*

9.   A   2   *1 mark*
        B   −4   *1 mark*

                                          Total = 20 marks

In order to proceed to the next programme you should have obtained at least:

> Question 4   (objective 1)   2/3   (Frames 4.9–4.23)
> Question 5   (objective 2)   3/4   (Frames 4.20–4.29)
> Question 6   (objective 3)   1/1   (Frame 4.29)
> Question 7   (objective 3) ⎫
> Question 8   (objective 3) ⎬ 5/6   (Frames 4.29–4.39)
> Question 9   (objective 3) ⎭

If you have obtained less than these scores you should return to the frames shown and ask somebody to set you some questions comparable to those you got wrong.

Matrices

Revision Notes

A matrix is an array of numbers, containing any number of rows and any number of columns. Unlike a determinant, it does not have a numerical value.

Two matrices (X) and (Y) can be combined in that order if the number of columns in (X) equals the number of rows in (Y). If this condition holds, the matrices are said to be conformable.

Combination of matrices is effected by working along the rows of the first matrix and down the columns of the second. An element in the rth row and the cth column of the product is formed by multiplying together the elements from the rth row of the first matrix and the cth column of the second and summing the products, e.g.

$$\begin{pmatrix} 1 & 4 \\ 6 & 8 \end{pmatrix} \begin{pmatrix} 2 & 3 \\ 5 & 7 \end{pmatrix} = \begin{pmatrix} 1 \times 2 + 4 \times 5 & 1 \times 3 + 4 \times 7 \\ 6 \times 2 + 8 \times 5 & 6 \times 3 + 8 \times 7 \end{pmatrix} = \begin{pmatrix} 22 & 31 \\ 52 & 74 \end{pmatrix}$$

A symmetry (or other) operation converts a set of vectors into a new set of vectors. If the original and the new set are written as column matrices, the operation can be represented by the square matrix which interconverts the two.

The character of a square matrix is the sum of the numbers on its principal diagonal.

For a matrix representing an operation the character is equal to the extent to which the basis vectors are converted to themselves by the operation (N.B. this may be a negative extent if directions are reversed.)

Degenerate Representations

Objective

After completing this programme you should be able to find the characters of a set of representations generated by using a set of degenerate vectors as a basis.

This objective is tested at the end of the programme.

Assumed Knowledge

A knowledge of the contents of the earlier programmes is assumed.

Note

This programme is the last one before the ones which deal with the applications of molecular symmetry. In many ways it seeks to link together the contents of the earlier programmes rather than to introduce radically new material.

Degenerate Representations

5.1   What are the point groups of the following?

A. CH_4 B. Benzene   C.   D. $CHCl_3$

5.2   A. T_d
        B. D_{6h} (Programme 2)
        C. D_{2h}
        D. C_{3v}

The character table for the C_{2h} point group is (in part):

C_{2h}	E	C_2	i	$\sigma_h(xy)$
A_g	1	1	1	1
B_g	1	-1	1	-1
A_u	1	1	-1	-1
B_u	1	-1	-1	1

N.B. z is vertical

Decide whether the x direction is symmetric or antisymmetric to the four group operations, and hence decide the symmetry species to which x belongs.

5.3   Symmetric to E and $\sigma(xy)$
        Antisymmetric to C_2 and i
            $\therefore$ x belongs to the B_u representation (Programme 3).
        Use the four arrows shown as a basis for generating a matrix to represent the operations i, $\sigma(xy)$ and $C_2(x)$ on the oxalate ion.
        Find the character of each matrix.

5.4   i: $\begin{pmatrix} 0 & 0 & 1 & 0 \\ 0 & 0 & 0 & 1 \\ 1 & 0 & 0 & 0 \\ 0 & 1 & 0 & 0 \end{pmatrix} \begin{pmatrix} a_1 \\ a_2 \\ a_3 \\ a_4 \end{pmatrix} = \begin{pmatrix} a_3 \\ a_4 \\ a_1 \\ a_2 \end{pmatrix}$ i.e. new a_1 = old a_3 etc.
Character = 0

$\sigma(xy)$: $\begin{pmatrix} 1 & 0 & 0 & 0 \\ 0 & 1 & 0 & 0 \\ 0 & 0 & 1 & 0 \\ 0 & 0 & 0 & 1 \end{pmatrix} \begin{pmatrix} a_1 \\ a_2 \\ a_3 \\ a_4 \end{pmatrix} = \begin{pmatrix} a_1 \\ a_2 \\ a_3 \\ a_4 \end{pmatrix}$ Character = 4

$C_2(x)$: $\begin{pmatrix} 0 & 1 & 0 & 0 \\ 1 & 0 & 0 & 0 \\ 0 & 0 & 0 & 1 \\ 0 & 0 & 1 & 0 \end{pmatrix} \begin{pmatrix} a_1 \\ a_2 \\ a_3 \\ a_4 \end{pmatrix} = \begin{pmatrix} a_2 \\ a_1 \\ a_4 \\ a_3 \end{pmatrix}$ Character = 0
(Programme 4)

If you have got these questions substantially correct, you can proceed with this programme; if not, you should return to the appropriate earlier programme to make good any deficiency.

We left the programme on non-degenerate representations at the point where we were considering the species to which x and y belonged in C_{4v} symmetry.

If we take the ion:

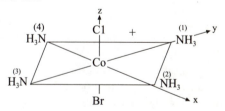

which has C_{4v} symmetry, and consider the effect of the group operations on a directional property such as a vector in the x direction, we find that x and y are interconverted by some of the group operations.

The group operations are   E, $2C_4$, $C_2 (=C_4^2)$, $2\sigma_v$, $2\sigma_d$, where each σ_v includes either the x or y axis and each σ_d lies between the axes. Which of the operations cause "mixing" of arrows along the x and y directions?

5.5 $2C_4$, $2\sigma_d$

What effect does a clockwise C_4 have on the NH_3 molecules on the x and y axes?

5.6   new 'x – NH_3' =   old 'y – NH_3'   (NH_3 numbered (1))
new 'y – NH_3' = –old 'x – NH_3'   (NH_3 numbered (4))

What about an anticlockwise C_4? (This is the same as C_4^3.)

5.7   new 'x – NH_3' = –old 'y – NH_3' NH_3 (3)
new 'y – NH_3' =   old 'x – NH_3' NH_3 (2)

Write down the two matrices which represent these two transformations.

What are their characters?

5.8 C_4:
$$\begin{pmatrix} 0 & 1 \\ -1 & 0 \end{pmatrix}\begin{pmatrix} x \\ y \end{pmatrix} = \begin{pmatrix} y \\ -x \end{pmatrix}$$
$$\begin{pmatrix} 0 & -1 \\ 1 & 0 \end{pmatrix}\begin{pmatrix} x \\ y \end{pmatrix} = \begin{pmatrix} -y \\ x \end{pmatrix}$$

Character = 0 in both cases

You should now realise why we use the term CHARACTER TABLE. The numbers are the CHARACTERS of the matrices which represent the group operations. In our simple examples of non-degenerate representations the matrices were all single numbers and the number was the same as the character of the matrix. Many of the theorems of Group Theory only involve the characters of the matrix representations of operations, so these are all that are included in the character table. Operations are grouped together in classes because all operations in the same class can be represented by matrices of the same character. (For a treatment of classes see Programme 2, frames 2.35–2.40.)

Use the x and y directions as a basis for representations of the two reflections σ_d Use the following convention:

(z is vertical)

5.9 σ_d $\begin{pmatrix} 0 & 1 \\ 1 & 0 \end{pmatrix}$ new x = old y

                                    new y = old x

       σ_d' $\begin{pmatrix} 0 & -1 \\ -1 & 0 \end{pmatrix}$ new x = −old y

                                    new y = −old x

In both cases the character is zero. This does not prove that the two operations are in the same class, but if they are in the same class, the characters of the two matrices must be equal.

Construct the matrices to represent the two σ_v operations, $\sigma(xz)$ and $\sigma(yz)$

5.10 $\sigma(xz)$: $\begin{pmatrix} 1 & 0 \\ 0 & -1 \end{pmatrix}$ $\sigma(yz)$: $\begin{pmatrix} -1 & 0 \\ 0 & 1 \end{pmatrix}$

In both cases the character is zero, but the σ_v operations are not in the same class as the σ_d operations because there are other representations where they have different characters.

What effect does the operation C_4 have on the total energy of a C_{4v} molecule?

5.11   None at all. It is a symmetry operation, so leaves the molecule indistinguishable.

We have seen that directional properties along x and y are interconverted by C_4 (e.g. p_x and p_y orbitals), so what can we say about the relative energies of p_x and p_y orbitals if they can be interconverted by a symmetry operation?

5.12   They must be identical, i.e., degenerate.

This was just a short reminder of something we have met already, and is the reason why the representation to which x and y both belong in C_{4v} is termed a DEGENERATE REPRESENTATION.

Use the transformation properties of the x and y axes to construct the matrices which represent all the operations of the C_{4v} group, namely
E, C_4, C_4^3, C_2 $(=C_4^2)$, $\sigma_v(xz)$, $\sigma_v(yz)$, σ_d, and σ_d'

5.13   E C_4 C_2 $\sigma_v(xz)$ σ_d

$$\begin{pmatrix} 1 & 0 \\ 0 & 1 \end{pmatrix} \begin{pmatrix} 0 & 1 \\ -1 & 0 \end{pmatrix} \begin{pmatrix} -1 & 0 \\ 0 & -1 \end{pmatrix} \begin{pmatrix} 1 & 0 \\ 0 & -1 \end{pmatrix} \begin{pmatrix} 0 & 1 \\ 1 & 0 \end{pmatrix}$$

C_4^3 $\sigma_v(yz)$ σ_d'

$$\begin{pmatrix} 0 & -1 \\ 1 & 0 \end{pmatrix} \qquad \begin{pmatrix} -1 & 0 \\ 0 & 1 \end{pmatrix} \begin{pmatrix} 0 & -1 \\ -1 & 0 \end{pmatrix}$$

Write down the group operations, and under each operation write the character of the matrix representing the operation. The result should be a row of the C_{4v} character table, i.e. the species to which both x and y belong.

5.14   E $2C_4$ C_2 $2\sigma_v$ $2\sigma_d$ (Note the grouping
        2   0   -2   0   0   into classes.)

This is labelled the E representation (do not confuse it with the identity element). We can now think a little about the meaning of some of the labels used for symmetry species – A and B both refer to 1-degenerate representations, E to a 2-degenerate representation, where e.g., x and y are mixed, and T refers to a 3-degenerate representation where e.g., x, y and z are all mixed.

The matrix representing the identity must combine with another matrix to leave it unchanged. For a 1-degenerate representation the identity matrix is (1) i.e., (1) (x) = (x). What is the square matrix (M) which represents the identity in a 2-degenerate representation? i.e., $(M) \begin{pmatrix} x \\ y \end{pmatrix} = \begin{pmatrix} x \\ y \end{pmatrix}$. What is its character?

5.15 $(M) = \begin{pmatrix} 1 & 0 \\ 0 & 1 \end{pmatrix}$ character = 2 i.e. $\begin{pmatrix} 1 & 0 \\ 0 & 1 \end{pmatrix} \begin{pmatrix} x \\ y \end{pmatrix} = \begin{pmatrix} x \\ y \end{pmatrix}$

What is the identity matrix in a 3-degenerate representation? What is its character?

5.16 $\begin{pmatrix} 1 & 0 & 0 \\ 0 & 1 & 0 \\ 0 & 0 & 1 \end{pmatrix}$ character $= 3$

i.e. $\begin{pmatrix} 1 & 0 & 0 \\ 0 & 1 & 0 \\ 0 & 0 & 1 \end{pmatrix} \begin{pmatrix} x \\ y \\ z \end{pmatrix} = \begin{pmatrix} x \\ y \\ z \end{pmatrix}$

We now have a quick and easy way of finding the degeneracy of a representation directly from the character table. Can you see what it is?

5.17 The degeneracy equals the character of the matrix representing the identity.

In the C_{4v} character table, x, y, xz, yz, rotation about x and rotation about y all belong to the E representation. They are not all mixed, however, by the group operations (we can obviously not mix an x direction with a rotation). In the character table, therefore, they are grouped together in brackets according to the way they mix, e.g.

C_{4v}	E	$2C_4$	C_2	$2\sigma_v$	$2\sigma_d$	
E	2	0	-2	0	0	$(x, y)(R_x, R_y)(xz, yz)$

This tells us that xz and yz are degenerate with each other in this symmetry, but not with x or y which are, however, degenerate with each other.

In frames 5.2 and 5.3 we saw that x belongs to the B_u representation of C_{2h}.

Decide whether the y direction is symmetric or antisymmetric to the four group operations of C_{2h} and hence decide the symmetry species to which y belongs.

5.18 B_u i.e. $\begin{aligned} Ey &= y \\ C_2y &= -y \\ iy &= -y \\ \sigma(xy)y &= y \end{aligned}$

Thus x and y both belong to the same representation of C_{2h}. Does this necessarily mean they are degenerate?

5.19   No, because the group operations do not interconvert x and
        y, they merely happen to belong to the same representation.
        This sort of thing happens a lot because there are many
        directional properties, but only a limited number of irredu-
        cible representations. In the character table for C_{2h} x and y
        are put on the same line *but are not bracketed together* e.g.

C_{2h}	E	C_2	i	σ_h	
B_u	1	-1	-1	1	x, y

        Let us now return to our matrix representations of C_{4v}. We
        have seen in an earlier programme that representations are
        called representations for two reasons:

        i. They represent the effect of the group operations on
            certain directional properties.
        ii. Can you remember the second reason (about combina-
            tion)?

5.20   They combine together in the same way as the group opera-
        tions. Let us check this for a few of the operations of C_{4v}.

        What is the effect of applying C_4 clockwise about z, followed
        by σ_d on the point A? (Call the new point A$'$, and decide
        which single operation would take A to A$'$.)

        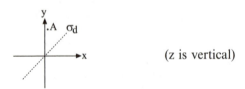

        (z is vertical)

5.21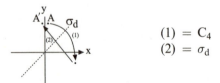

        (1) = C_4
        (2) = σ_d

A is taken to A$'$ by $\sigma(yz)$ i.e. $\sigma_d C_4 = \sigma(yz)$ (remember we write $\sigma_d C_4$
to mean C_4 followed by σ_d). Multiply together the two matrices (see
frame 5.13) representing C_4 and σ_d in the order $\sigma_d C_4$ to see if they
give the matrix representing $\sigma(yz)$.

5.22

$$\begin{pmatrix} 0 & 1 \\ 1 & 0 \end{pmatrix}\begin{pmatrix} 0 & 1 \\ -1 & 0 \end{pmatrix} = \begin{pmatrix} -1 & 0 \\ 0 & 1 \end{pmatrix}$$

$\qquad \sigma_d \quad C_4 \qquad = \sigma(yz)$

Do C_4 and σ_d commute?

5.22A. If they commute, then $\sigma_d C_4 = C_4 \sigma_d$ remember?

5.23 They do not commute,

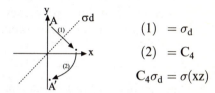

$$(1) \ = \sigma_d$$
$$(2) \ = C_4$$
$$C_4 \sigma_d = \sigma(xz)$$

Does this agree with the matrix representation?

5.24 Of course,

$$\begin{pmatrix} 0 & 1 \\ -1 & 0 \end{pmatrix}\begin{pmatrix} 0 & 1 \\ 1 & 0 \end{pmatrix} = \begin{pmatrix} 1 & 0 \\ 0 & -1 \end{pmatrix}$$

$\qquad C_4 \qquad \sigma_d \quad = \sigma(xz)$

Try the same thing for the two operations C_4^3 and $\sigma(yz)$.

5.25

$$\begin{pmatrix} -1 & 0 \\ 0 & 1 \end{pmatrix}\begin{pmatrix} 0 & -1 \\ 1 & 0 \end{pmatrix} = \begin{pmatrix} 0 & 1 \\ 1 & 0 \end{pmatrix} \Bigg| \begin{pmatrix} 0 & -1 \\ 1 & 0 \end{pmatrix}\begin{pmatrix} -1 & 0 \\ 0 & 1 \end{pmatrix} = \begin{pmatrix} 0 & -1 \\ -1 & 0 \end{pmatrix}$$

$$\sigma(yz) \quad C_4^3 \quad = \quad \sigma_d \quad \Bigg| \quad C_4^3 \quad \sigma(yz) \quad = \quad \sigma_d'$$

$$(1) = C_4^3 \qquad\qquad\qquad (1) = \sigma(yz)$$
$$(2) = \sigma(yz) \qquad\qquad\qquad (2) = C_4^3$$

You could, if you wish, set up the whole 8×8 multiplication table for the group, using the E representation, but it is not really worth it – the representation is a genuine one, and any combination of symmetry operations is paralleled by the corresponding combination of matrices, taken in the correct order.

What is the point group of the molecule CH_4?

5.26 T_d, the tetrahedral group. Find its character table in the book of tables. Let us set up a representation of T_d using as a basis the four C—H bonds of methane:

$$\begin{array}{c} H_{(1)} \\ | \\ C \\ \diagdown \\ H_{(4)} \quad H_{(3)} \quad H_{(2)} \end{array}$$

What is the order of the T_d group?

5.26A   The order is the number of operations in the group, remember? Now count them up, using the character table.

5.27   24

A complete set of representations will therefore consist of twenty four 4×4 matrices. This is a bit much but we can simplify the problem in two ways.

What property of a square matrix can we often use instead of the full matrix?

5.28   Its character.

The eight C_3 operations are all in the same class. What does this tell you about the characters of all the eight matrices representing the C_3 operations?

5.29   The characters are all the same, because all eight operations are in the same class.

We need, therefore, consider only one representative operation in each class. Let us take the clockwise rotation about bond 1. What effect does this have on each bond? i.e. what bond moves to position (4) to become the new bond (4) etc.?

5.30   New bond 1 = old bond 1
New bond 2 = old bond 3
New bond 3 = old bond 4
New bond 4 = old bond 2

Write this in matrix form and find the character.

5.31 $$\begin{pmatrix} 1 & 0 & 0 & 0 \\ 0 & 0 & 1 & 0 \\ 0 & 0 & 0 & 1 \\ 0 & 1 & 0 & 0 \end{pmatrix} \begin{pmatrix} B_1 \\ B_2 \\ B_3 \\ B_4 \end{pmatrix} = \begin{pmatrix} B_1 \\ B_3 \\ B_4 \\ B_2 \end{pmatrix} \qquad \text{Character} = 1$$

Can you remember a quick way to find the character of such a matrix from the information in frame 5.30 above?

5.32   The character equals the number of bonds unshifted by the operation, i.e., the character is only influenced by the extent to which a bond is transformed to itself. This is the second of our simplifications.

How many bonds are *un*shifted by:

   i. The identity?
   ii. One of the three C_2 operations?

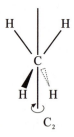

5.33   i. Four
           ii None

Hence what are the characters of the representations of E and C_2 using the four-bond basis?

5.34   4 and 0 respectively.

We have already seen that the character of the matrix representing C_3 is 1. How many bonds are left unshifted by:

   i. One of the six S_4 operations (S_4 axis is colinear with C_2)?
   ii. One of the six planes (the plane of the paper in frame 5.32 above)?

Hence complete the representation:

	E	$8C_3$	$3C_2$	$6S_4$	$6\sigma_d$
Γ_1	4	1	0		

5.35

	E	$8C_3$	$3C_2$	$6S_4$	$6\sigma_d$
Γ_1	4	1	0	0	2

Is this a representation in the T_d character table?

5.36   No. It is a reducible representation (strictly a set of characters of a reducible representation).

Reduce it, then, using the T_d character table.

5.37 $\Gamma_1 = A_1 + T_2$ (Programme 3)

If we look at the character table, we can see that the p_x, p_y and p_z orbitals belong to T_2. Which orbital do you think belongs to the totally symmetric representation A_l, i.e., what type of orbital is unaffected by any symmetry operation?

5.38   An s orbital which is spherically symmetrical and hence symmetric to all operations of any group.

We have found that our reducible representation contains the irreducible representations to which s and the three p orbitals belong. Thus if we combine an s and three p orbitals, we will get a set of hybrid orbitals pointing towards the corners of a tetrahedron, i.e., an sp^3 set is a set of tetrahedral hybrids – symmetry theory is producing results at last!

The set of p orbitals is not the only set belonging to T_2. What is the other set?

5.39   The d orbitals d_{xy} d_{xz} d_{yz}

Thus from symmetry alone we cannot distinguish a set of sp^3 hybrids from a set of sd^3 hybrids. This is another example of how symmetry will give us so much information but no more. We need further calculations to tell us that sp^3 hybrids are likely to be important in CH_4, but sd^3 hybrids are likely to be more important in MnO_4^-.

This programme has been partly a linking together of a lot of the previous work, but you should also be able to find the characters of a set of representations generated by using a set of degenerate vectors as a basis. The following test will show you how well you have learned this.

Degenerate Representations Test

1. A. Write out the characters of the representation of C_{4h} using x and anything degenerate with x as basis. The group operations are given below, all axes are vertical and colinear,

    $$E \quad C_4 \quad C_2(=C_4{}^2) \quad C_4{}^3 \quad i \quad S_4{}^3 \quad \sigma_h \quad S_4$$

    B. With what, if anything, is x degenerate?

2. A. As question 1A using a d_{xz} orbital and anything degenerate with it as basis.
    B. With what, if anything, is d_{xz} degenerate?

3. A. As question 1A using x and anything degenerate with it as a basis for D_{4h}. The group operations are:

    $$E \quad 2C_4 \quad C_2 \quad 2C_2{}' \quad 2C_2{}'' \quad i \quad 2S_4 \quad \sigma_h \quad 2\sigma_v \quad 2\sigma_d$$

    ($2C_4$, C_2 and $2S_4$ are vertical. $2C_2{}'$ and $2\sigma_v$ include an x or y axis $2C_2{}''$ and $2\sigma_d$ lie between the x and y axes).

    B. With what, if anything, is x degenerate?

4. A. As question 1A using a d_{xz} orbital and anything degenerate with it as a basis for D_{4h}
    B. With what, if anything, is d_{xz} degenerate?

Answers

One mark for each underlined answer you get right.

	E	C_4	C_2	$C_4{}^3$	i	$S_4{}^3$	σ_h	S_4
1.A.	2	0	−2	0	−2	0	2	0
2.A.	2	0	−2	0	2	0	−2	0

1.B.	y	—	*1 mark*
2.B.	yz	—	*1 mark*

	E	$2C_4$	C_2	$2C_2{}'$	$2C_2{}''$	i	$2S_4$	σ_h	$2\sigma_v$	$2\sigma_d$
3.A.	2	0	−2	0	0	−2	0	2	0	0
4.A.	2	0	−2	0	0	2	0	−2	0	0

3.B.	y	—	*1 mark*
4.B.	yz	—	*1 mark*

Total   40 marks

The test score on this programme is very much less critical than the others, but a score below 30 indicates that you have not really understood the material very well. The average score of the students who tested the programme before publication of the first edition was 36.

Degenerate Representations

Revision Notes

If a group includes a proper axis with an order of 3 or more, the application of some symmetry operations causes one directional property to be converted to another. If there is an energy associated with the directional properties, e.g. the energy of p_x and p_y orbitals, these energies must be identical, i.e. symmetry tells us directly that two directional properties which are mixed by symmetry must be degenerate.

If two directional properties are mixed by symmetry operations, the operations can only be represented by matrices, whose character appears in the character table. The directional properties mixed by symmetry operations are bracketed together in the character table, e.g. (x, y); (xz, yz) etc.

The degeneracy of a degenerate representation is equal to the character of the identity matrix.

> One-degenerate representations are labelled A or B.
> Two-degenerate representations are labelled E.
> Three-degenerate representations are labelled T.

Applications to Chemical Bonding

Objectives

After completing this programme you should be able to:

1. Find sets of hybrid orbitals with given directional properties.
2. Determine the orbitals suitable for π-bonding in a molecule.
3. Find the symmetries of LCAO molecular orbitals.
4. Construct simple MO correlation diagrams.

All four objectives are tested at the end of the programme.

Assumed Knowledge

A knowledge of the contents of Programmes 1–5 is assumed.

Applications to Chemical Bonding

6.1   If you have worked through, and understood, the five preceding programmes on Group Theory, you should now be ready to tackle either of the programmes on applications. If not, you should go back and be sure you understand the theory before trying to apply it.

We will look at four applications of Group Theory in this programme:

i.   Construction of hybrid orbitals (frames 6.2–6.10).
ii.   Finding orbitals suitable for π-bonding (frames 6.10–6.17).
iii.   Determination of the symmetry of LCAO molecular orbitals (frames 6.17–6.22).
iv.   Construction of qualitative molecular orbital correlation diagrams (frames 6.22–6.36).

(A dashed line separates each section of the programme.)

In most cases the use of Group Theory can be summarised in three rules:

i.   Use an appropriate *basis* to generate a *reducible representation* of the *point group*.
ii.   *Reduce* this representation to its constituent *irreducible representations*.
iii.   Interpret the results.

(The construction of correlation diagrams is a little more complicated than this.)

Do you understand all the italicised terms in the above rules?

6.2   If there are any of these terms you do not understand, return to the appropriate earlier programme:

Basis: Programme 4 frames 4.33–4.39.
Reducible Representation: Programme 3 frames 3.17–3.25.
Point Group: Programme 2 frames 2.1–2.24.
Reduce: Programme 3 frames 3.18–3.25.
Irreducible Representation: Programmes 3 and 5.

We will start with the construction of a set of hybrid orbitals. We have already seen in the previous programme (frames 5.26–5.39) how to do this for a tetrahedral set, so for this example we will use a trigonal plane shape, and find which orbitals can be hybridised to produce a set of three trigonal planar σ bonds.

What is the point group whose character table we shall need to use?

6.3 D_{3h}, the point group of a trigonal planar molecule like BCl_3. What set of vectors could represent a set of trigonal planar bonds?

6.4   A set of three vectors as follows:

We can use this set of vectors as a basis to generate a reducible representation of the D_{3h} point group.
    The operations of D_{3h} are:

E $2C_3$ $3C_2$ σ_h $2S_3$ $3\sigma_v$

Can you remember the simple way of finding the character of a matrix representing a particular operation?

6.5   The character equals the extent to which the vectors are transformed to themselves, or in this simple case the number of vectors unshifted by the operation.

Use this simplification to write down the characters of the representations of E, C_3 and C_2. The answer gives the characters and the full matrix equations.

6.6   E,   Character $= 3$ $\begin{pmatrix} 1 & 0 & 0 \\ 0 & 1 & 0 \\ 0 & 0 & 1 \end{pmatrix} \begin{pmatrix} a_1 \\ a_2 \\ a_3 \end{pmatrix} = \begin{pmatrix} a_1 \\ a_2 \\ a_3 \end{pmatrix}$

$$C_3, \text{(clockwise)}, \; \chi = 0 \begin{pmatrix} 0 & 1 & 0 \\ 0 & 0 & 1 \\ 1 & 0 & 0 \end{pmatrix} \begin{pmatrix} a_1 \\ a_2 \\ a_3 \end{pmatrix} = \begin{pmatrix} a_2 \\ a_3 \\ a_1 \end{pmatrix}$$

$$C_2, \text{(through } a_1), \; \chi = 1 \begin{pmatrix} 1 & 0 & 0 \\ 0 & 0 & 1 \\ 0 & 1 & 0 \end{pmatrix} \begin{pmatrix} a_1 \\ a_2 \\ a_3 \end{pmatrix} = \begin{pmatrix} a_1 \\ a_3 \\ a_2 \end{pmatrix}$$

Now go on and find the characters of the representations of the other operations.

6.7 $\sigma_h, \chi = 3$ All vectors remain unshifted.

$S_3, \chi = 0$ All vectors are shifted.

$\sigma_v, \chi = 1$ The plane passes through one arrow and leaves it unshifted.

The complete set of characters is thus:

D_{3h}	E	$2C_3$	$3C_2$	σ_h	$2S_3$	$3\sigma_v$
Γ_1	3	0	1	3	0	1

This is a set of characters of a reducible representation of D_{3h}. In previous programmes we loosely called such a set of numbers a reducible representation. It is vital to the use of Group Theory that you should be able to reduce such a representation, so use the character table to reduce it.

6.8 $\Gamma_1 = A_1' + E'$

e.g. number of $A_1' = \frac{1}{12}(3 + 0 + 3 + 3 + 0 + 3) = 1$

number of $A_2' = \frac{1}{12}(3 + 0 - 3 + 3 + 0 - 3) = 0$

number of $E' = \frac{1}{12}(6 + 0 + 0 + 6 + 0 + 0) = 1$ etc.

If you have not achieved this result, it is essential that you return to the reduction formula in Programme 3 frame 18 to refresh your memory.

Look at the right-hand side of the D_{3h} character table to decide which orbitals belong to the symmetry species A_1' and E'.

6.9 A_1' includes either the d_{z^2} or the spherically symmetrical s
        orbital. E' includes p_x and p_y together or $d_{x^2-y^2}$ and d_{xy}
        together, i.e. we know that p_x and p_y are degenerate, as are
        $d_{x^2-y^2}$ and d_{xy} because they are bracketed together in the two-
        degenerate E' representation.

        What, then is the most likely set of hybrids to form a trigonal
        set of bonds in a *first row* element like boron?

6.10   s p_x p_y i.e. an sp^2 set. The plane is conventionally taken to be
        the xy plane, z is vertical.

        We have now been through all the stages outlined in frame
        6.1.

        i.   The *basis* of our reducible representation was a set of
              vectors representing the bonds.
        ii.   We *reduced* it to $A_1' + E'$.
        iii. We *interpreted* the results to mean hybridisation of s, p_x
              and p_y orbitals.

        The most crucial step in this process is the first one. The
        correct choice of basis is vital. It must reflect the question
        we are asking the theory to answer. Get the basis right and
        everything else follows easily.

        Note that there is no reason why hybridisation of d_{z^2}, $d_{x^2-y^2}$,
        and d_{xy} should not be equally acceptable on symmetry
        grounds – Group Theory will only take us so far in a calcula-
        tion, we have then to do further calculations or at least select
        the most reasonable of the alternatives given by symmetry.

        -

        Let us now see which orbitals would be suitable for π-bond-
        ing in a D_{3h} molecule. Remember that a π-bond has a wave
        function whose sign differs in the two lobes:

        Draw an arrow which could represent the symmetry proper-
        ties of this orbital. (Call the point of the arrow the positive
        end.)

6.11 · ↑ · represents the symmetry of the π-orbital. Remember that each pair of atoms could be linked by two π-bonds at right angles, and draw a suitable set of six arrows to act as a basis for a representation of the possible if π-bonds in a D_{3h} molecule of formula AB_3.

6.12

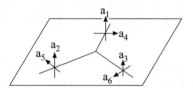

These are in two sets, a_1, a_2, a_3 – the "out of plane" set, and a_4, a_5, a_6 – the "in plane" set. The two sets will clearly not be mixed by any of the group operations, so we can consider each separately.

Consider the extent to which a_1, a_2, and a_3 are converted to themselves by the group operations (remember that upwards is the positive direction), and hence write down the characters of the representation generated by the "out of plane" set of arrows. The group operations are:

| E | $2C_3$ | $3C_2$ | σ_h | $2S_3$ | $3\sigma_v$ |

6.13

	E	$2C_3$	$3C_2$	σ_h	$2S_3$	$3\sigma_v$
Γ_2	3	0	-1	-3	0	1

Do the same thing for the "in plane" set.

6.14

	E	$2C_3$	$3C_2$	σ_h	$2S_3$	$3\sigma_v$
Γ_3	3	0	-1	3	0	-1

Reduce Γ_2 and Γ_3

6.15 Γ_2 (out of plane) $= A_2'' + E''$
Γ_3 (in plane) $= A_2' + E'$

Look at the character table to decide which orbitals are suitable for π-bonding of the two types.

6.16   Out of plane: p_z (d_{xz}, d_{yz}) together.

In plane: (p_x, p_y) together or ($d_{x^2-y^2}$, d_{xy}) together.

(N.B. there is no orbital of symmetry A_2'.)

For a first row element such as boron, there are no energetically available d orbitals. The p_x and p_y orbitals, although π-orbitals in a local diatomic sense, are involved in σ-bonding in a molecule like BCl_3 (frame 6.10), so we are left with only one orbital which is a true π-orbital with respect to the whole molecular plane.

Which orbital is this?

6.17   The p_z orbital e.g. BCl_3:

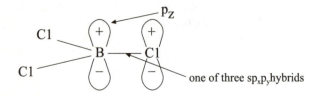

Again, we have used the same procedure to solve the problem. A different *basis* was used for this particular example but the *reduction* process should by now be second nature and the *interpretation* of the results needed a little care. The crucial step, however, was again the selection of the correct basis to reflect the orbitals we wished to find.

The result we obtained using group theory suggests that in BCl_3, and indeed in the other boron trihalides, there would be some π-bonding. This would involve electrons being given from a filled chlorine orbital to the empty p_z orbital of boron. Some aspects of the chemistry of the boron halides provide strong evidence for the existence of this bonding. Many advanced inorganic chemistry texts include a discussion of these aspects.

We will now turn to the question of the symmetries of LCAO molecular orbitals. These are made by taking linear combinations of the constituent atomic orbitals (LCAO), and the atomic orbitals form a convenient basis for the reducible representation of the group. We will again use a D_{3h} molecule as an example, and will find the symmetries of the π-molecular orbitals of the radical:

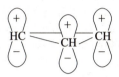

Use the transformation properties of these three atomic orbitals to find the characters of a representation of D_{3h}:

D_{3h}	E	$2C_3$	$3C_2$	σ_h	$2S_3$	$3\sigma_v$

6.18

D_{3h}	E	$2C_3$	$3C_2$	σ_h	$2S_3$	$3\sigma_v$
Γ_4	3	0	-1	-3	0	1

Reduce this representation.

6.19 $\Gamma_4 = A_2'' + E''$

i.e. Γ_4 is the same as Γ_2, formed from the out of plane π-bonds of a molecule like BCl_3. (This is a result you may have expected from a consideration of the two bases used.) This result tells us that the molecular orbitals consist of a doubly degenerate pair (E'') and one singly degenerate orbital (A_2''). The result tells us nothing about the energy difference between the A_2'' and the E'' orbitals nor does it tell us anything of the absolute energies of any of the orbitals.

The energies of the orbitals can be readily calculated using Huckel molecular orbital theory in terms of the energies α and β. Details of the theory are outside the scope of this book, but α and β are both negative amounts of energy so that an orbital of energy $(\alpha + \beta)$ is a very low energy orbital. Huckel theory applied to the cyclopropenyl ion gives the orbital energies as $(\alpha + 2\beta)$, $(\alpha - \beta)$ and $(\alpha - \beta)$, i.e. a single orbital (A_2'') and a degenerate pair (E''). We can follow the same procedure for the hypothetical molecule cyclobutadiene:

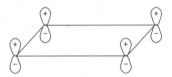

What is the point group of this molecule?

6.20 D_{4h} The group operations are:

E $2C_4$ $C_2(= C_4{}^2)$ $2C_2'$ $2C_2''$ i $2S_4$ σ_h $2\sigma_v$ $2\sigma_d$

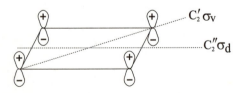

Write down the reducible representation of D_{4h} formed by using the four atomic p orbitals as a basis.

6.21

D_{4h}	E	$2C_4$	C_2	$2C_2'$	$2C_2''$	i	$2S_4$	σ_h	$2\sigma_v$	$2\sigma_d$
Γ_5	4	0	0	-2	0	0	0	-4	2	0

Use the D_{4h} character table to reduce this representation.

6.22 $\Gamma_5 = E_g + A_{2u} + B_{2u}$

i.e. there are two singly degenerate orbitals and a degenerate pair. This again agrees with simple calculations which show the energies to be $(\alpha + 2\beta)$, α (twice), $(\alpha - 2\beta)$. The E_g orbitals clearly have energy α, and the other two correspond to the singly degenerate ones.

Again a suitable choice of *basis* enabled us to generate a representation of the group to solve the problem.

In the final section of this programme we will consider the subject of molecular orbital correlation diagrams. These diagrams show the energies and symmetries of molecular orbitals and of the atomic orbitals from which they are constructed. As in other applications involving energy, symmetry considerations tell us nothing about energy differences – these have to be the subject of separate calculations. A knowledge of symmetry, however, does help when reading published accounts of molecular orbital calculations since orbitals are commonly labelled with their symmetry species.

A correlation diagram for water (C_{2v}) is shown below:

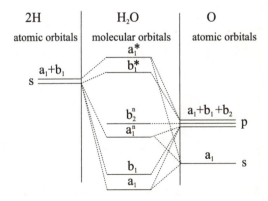

The energy levels on the outside of the diagram represent the s and p orbitals in the outer shell of the oxygen atom, and the s orbital of each hydrogen atom. We will now see how the symmetry labels are assigned, and the molecular orbitals in the centre of the diagram are derived.

Look at a C_{2v} character table and decide on the symmetry species of a p_x orbital of oxygen.

6.23 B_1 – the same as the x direction.

Hence the p_x orbital is labelled b_1, lower case letters being commonly use for particular orbitals.

Similarly decide on the labels of the s, p_y and p_z orbitals of the oxygen.

6.24   s is labelled a_1
       p_y is labelled b_2
       p_z is labelled a_1

These labels are included in the correlation diagram.

When we come to the two hydrogen atoms, it is necessary to consider the two ls orbitals.

Use the two ls orbitals ϕ_1 and ϕ_2 as the basis of a representation of the C_{2v} group:

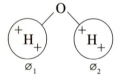

E C_2 σ σ'
σ is molecular plane

6.25

C_{2v}	E	C_2	σ	σ'
Γ_6	2	0	2	0

Reduce this representation.

6.26 $\Gamma_6 = A_1 + B_1$

The two linear combinations are therefore labelled a_1, and b_1 on the correlation diagram.

The actual wave functions of these two linear combinations are shown below:

$$\psi_1 = \frac{1}{\sqrt{2}}(\phi_1 + \phi_2) \qquad \psi_2 = \frac{1}{\sqrt{2}}(\phi_2 - \phi_1)$$

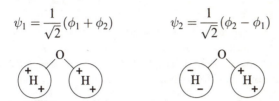

Use the transformation properties of ψ_1 and ψ_2 under the operations of the C_{2v} group to decide which is A_1 and which is B_1.

6.27 ψ_1 is symmetric to all the operations $\therefore$ it is A_1
ψ_2 is symmetric to E and σ
    antisymmetric to C_2 and σ' $\therefore$ it is B_1

Draw the p orbital of oxygen which belongs to the B_1 representation of C_{2v} i.e. has the same symmetry as ψ_2 above.

6.28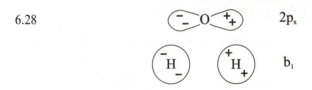

$2p_x$

b_1

An interaction can occur between orbitals of the central and outside atoms provided those orbitals have matching symmetry. In this case, we can add the two orbitals together to produce a low-energy bonding molecular orbital or subtract them to produce a high-energy antibonding orbital. Orbital combination of this type cannot, of course, change the total number of orbitals, so combining the two orbitals produces two molecular orbitals as a result.

Both of the resulting molecular orbitals have B_1 symmetry and are labelled b_1^* for the antibonding orbital.

Draw the b_1^* orbital obtained by subtracting the $2p_x$ orbital from the combination of hydrogen orbitals.

6.29

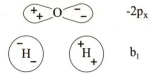

The crux of the symmetry aspect of molecular orbital theory is that atomic orbitals on different atoms will only interact if they belong to the same irreducible representation of the point group. In our water example, there is one orbital which does not match up with any from the other atom. Can you see which orbital this is?

6.30   The $2p_y$ orbital on oxygen labelled b_2.

This orbital does not intereact at all with the hydrogen orbitals – it remains non bonding, and is labelled on the correlation diagram b_2^n.

We have so far looked at orbitals of B_1 and B_2 symmetry. The only ones left are of A_1 symmetry. In this case there are two oxygen orbitals and only one from the combined ls orbitals of hydrogen. Calculations show that in this case there are three molecular orbitals, one bonding, one antibonding, and one non bonding. These are labelled on the diagram.

In general we cannot tell from symmetry arguments anything about the relative energies of orbitals. By their nature, however, bonding orbitals are of low energy; antibonding orbitals are of high energy and non bonding orbitals are in between. We can, therefore, draw a reasonable correlation diagram for the water molecule as shown below:

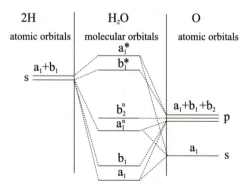

Our final job in describing the electronic structure of the water molecule is to put electrons into the molecular orbitals. How many electrons will there be from the 1s orbitals of two hydrogens and the 2s and 2p orbitals of oxygen?

6.31   Eight.   i.e.   one from each hydrogen
                         six from the oxygen

Put these into the molecular orbitals starting from the lowest energy orbital.

H_2O

——— a_1^*
——— b_1^*

——— b_2^n
——— a_1^n

——— b_1
——— a_1

6.32

——— a_1^* ⎫
——— b_1^* ⎬ anti bonding

⇅ b_2 ⎫
⇅ a_1 ⎬ non bonding

⇅ b_1 ⎫
⇅ a_1 ⎬ bonding

This description of the molecule puts two pairs of electrons in bonding orbitals and two into non bonding orbitals i.e. a very similar description to the valence bond description:

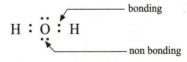

Finally, we will go through a slightly more complicated correlation diagram, that for the σ bonds in an octahedral complex ion like $[Co(NH_3)_6]^{3+}$. We shall need to consider the irreducible representations to which the 3d, 4s and 4p orbitals of cobalt belong. Look these up in the O_h character table.

6.33   3d: $(x^2 - y^2$ and $z^2)$ E_g
            $(xy, xz,$ and $yz)$ T_{2g}
   4s: A_{1g}
   4p: $(x, y,$ and $z)$ T_{1u}

The ligand orbitals which can form σ bonds can be represented by six arrows from the ligands to the metal:

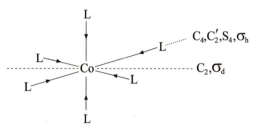

Try using this set of six arrows as a basis for a representation of O_h. This is quite difficult without some guidance so do not spend too long on it. The group operations are:

O_h E $8C_3$ $6C_2$ $6C_4$ $3C_2'(= C_4^2)$ i $6S_4$ $8S_6$ $3\sigma_h$ $6\sigma_d$

6.34

O_h	E	$8C_3$	$6C_2$	$6C_4$	$3C_2(=C_4^2)$	i	$6S_4$	$8S_6$	$3\sigma_h$	$6\sigma_d$
Γ_7	6	0	0	2	2	0	0	0	4	2

Reduce this reducible representation.

6.35 $\Gamma_7 = A_{1g} + E_g + T_{1u}$ We now have the start of our correlation diagram.

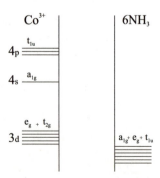

There is again one set of orbitals without any matching symmetry orbital on the other side of the diagram. Which is this?

6.36   The T_{2g} set of three metal ion orbitals.

These remain non bonding while in all other cases the orbitals combine to produce bonding and antibonding molecular orbitals:

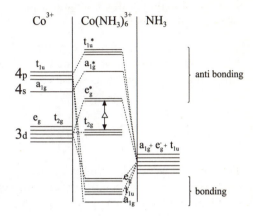

The complex $Co(NH_3)_6^{3+}$ has eighteen electrons in the orbitals under consideration. These will fill the molecular orbitals up to the non bonding t_{2g} level, giving six pairs of bonding electrons and six non bonding electrons which belong essentially to the metal. We can label the gap between the t_{2g} and e_g^* levels Δ and the picture is then remarkably similar to the ligand field theory picture of the bonding.

You should now be able to use Group Theory to find simple sets of hybrid orbitals, to determine the orbitals suitable for π-bonding in a molecule to find the symmetries of LCAO molecular orbitals and to construct simple MO correlation diagrams. The test overleaf consists of one problem on each of these applications.

Applications to Chemical Bonding Test

1. Find the hybrid orbitals of a central atom suitable for forming a set of square planar bonds. Use the D_{4h} character table.

2. Find the orbitals suitable for "out of plane" π-bonding in a square planar molecule.

3. Find the symmetries of the LCAO π-molecular orbitals of the open chain C_3 system: use the C_{2v} character table. How many different energy levels will there be in the system?

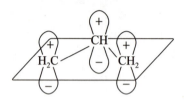

4. Set up the correlation diagram for the CH_4 molecule, Consider the 2s and 2p orbitals of carbon and the 1s orbital of each hydrogen atom.

Answers

1. Reducible representation:

D_{4h}	E	$2C_4$	C_2	$2C_2'$	$2C_2''$	i	$2S_4$	σ_h	$2\sigma_v$	$2\sigma_d$	
Γ_7	4	0	0	2	0	0	0	4	2	0	*2 marks*

This reduces to: $A_{1g} + B_{1g} + E_u$ *2 marks*

Suitable orbitals are: $A_{1g} - $ s or d_{z^2})
)
                                   $B_{1g} - d_{x^2-y^2}$)   *1 mark*
)
                                   $E_u - p_x$ and p_y together)

Hence a set of dsp^2 hybrid orbitals.

2. Reducible representation:

D_{4h}	E	$2C_4$	C_2	$2C_2'$	$2C_2''$	i	$2S_4$	σ_h	$2\sigma_v$	$2\sigma_d$	
Γ	4	0	0	-2	0	0	0	-4	2	0	*2 marks*

This reduces to: $E_g + A_{2u} + B_{2u}$ *2 marks*

Suitable orbitals are: $E_g - d_{xz}, d_{yz}$)
)
                                   $A_{2u} - p_z$)   *1 mark*
)
                                   $B_{2u} - $ none   )

3. Reducible representation:

C_{2v}	E	C_2	σ	σ'	
Γ	3	-1	-3	1	*1 mark*

This reduces to: $A_2 + 2B_2$ *1 mark*

i.e. 3 orbitals, all of different energy   *1 mark*

4.   Carbon orbitals: $A_1 + T_2$ *1 mark*

T_d	E	$8C_3$	$3C_2$	$6S_4$	$6\sigma_d$
1s of 4H	4	1	0	0	2

1 marks

This reduces to: $A_1 + T_2$ *1 mark*

Hence:

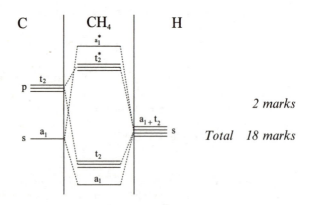

2 marks

Total   18 marks

Applications to Chemical Bonding

Revision Notes

The application of Group Theory to many chemical problems can be summarised in three rules:

i.   Use an appropriate *basis* to generate a reducible representation of the point group.
ii. *Reduce* this representation to its constituent irreducible representations.
iii. *Interpret* the results.

The initial choice of the basis is crucial. In essence this determines the question we are asking the theory to answer. If this is correct, the rest of the process follows virtually automatically.

The following applications require the bases shown:

i.   Hybrid orbitals – arrows representing the bonds.
ii. Orbitals suitable for π-bonding – arrows (two per pair of atoms) representing π-bonds.
iii. LCAO molecular orbitals – the constituent atomic orbitals.
iv. MO correlation diagrams – atomic orbitals of any central atom are allowed to interact with linear combinations of the orbitals of outer atoms which have the same symmetry.

Applications to Molecular Vibration

Objectives

After completing this programme you should be able to:

1. Find the symmetry species of the normal modes of vibration of a molecule of a given symmetry.
2. Find the number of infrared and Raman active vibrations in a molecule.
3. Find the number of active vibrations in a characteristic region of the infrared or Raman spectrum of a molecule.

All three objectives are tested at the end of the programme.

Assumed Knowledge

A knowledge of the contents of Programmes 1–5 is assumed. Some familiarity with vibrational spectroscopy will be found helpful.

Applications to Molecular Vibration

7.1   If you have worked through, and understood, Programmes 1
       to 5 on Group Theory, you should now be ready for this one.
       If not, you should go back and be sure you understand the
       underlying theory before trying to apply it.

       In this programme, we shall look at the use of Group Theory
       to find the symmetries of the vibrational modes of mole-
       cules, and we shall see which of the vibrations are observable
       in the infrared and Raman spectra. The programme is in
       three sections, separated by dashed lines.

       The use of Group Theory can be summarised in the follow-
       ing three rules:

       i.   Use an appropriate *basis* to generate a *reducible repre-
              sentation* of the *point group*.
       ii.   *Reduce* this representation to its constituent *irreducible
              representations*.
       iii.   Interpret the results.

       Do you understand all the italicised terms in the above rules?

7.2   If there are any of these terms which you do not understand,
       return to the appropriate earlier programme:

       Basis: Programme 4 frames 4.33–4.39
       Reducible Representation: Programme 3 frames 3.17–3.25
       Point Group: Programme 2 frames 2.1–2.24
       Reduce: Programme 3 frames 3.18–3.25
       Irreducible Representation: Programmes 3 and 5

       Group Theory can be an enormous help in deciding the
       infrared or Raman activity of different molecular vibrations,
       but before considering spectra we must look more generally
       at the subject of vibrations.

       Any movement of an atom in a molecule can be resolved into
       three components along the x, y, and z axes. If, therefore, there
       are n atoms in a molecule there are 3n possible movements of
       its atoms. Of these, 3 will be concerted movements of the whole
       molecule along the three co-ordinate axes, i.e. translations,
       and 3 (or 2 for a linear molecule) will be concerted rotations
       about the axes. The remaining $3n - 6$ (or $3n - 5$ for a linear
       molecule) must therefore be molecular vibrations.

       How many vibrations will there be for the molecule XeF_4?

7.3   9, i.e. there are 5 atoms and $3 \times 5 - 6 = 9$.

We can find the symmetries of all the possible molecular motions by using x, y, and z directions on each atom as a basis for a reducible representation of the group. For an n-atom molecule, this will produce a representation of order 3n, i.e. the character of the identity representation will be 3n, and all the matrices involved will be 3n × 3n matrices. This will obviously make it quite impracticable to set up the whole matrix for large molecules so we will need to use a quick means of finding the character of the matrix.

What is the quick way of finding the character of a matrix generated by any basis?

7.4   The character is equal to the extent to which the vectors in the basis are left unshifted by the operation.

Let us now use this procedure for the water molecule. The basis of the representation is the set of nine arrows:

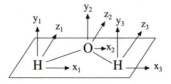

What is the point group of the water molecule, and what symmetry operations are there in the group? (Use the scheme in Programme 2 if you are not sure.)

7.5 C_{2v}

E C_2 σ σ'

Remember our quick way of finding the character of a matrix generated by a particular basis, and write down the characters of the 9×9 matrices representing E and C_2, using the nine-arrow basis shown. Remember that the arrows start at the atom, so could be reversed by some operations (i.e. give the contribution of -1 to the character).

7.6   E: $\chi = 9$ (all arrows unshifted)

C_2: $\chi = -1$ (all arrows on atoms 1 and 3 are shifted,

x_2 becomes $-x_2$

y_2 becomes $-y_2$

z_2 becomes $+z_2$)

Work out the characters of the representations of σ and σ' in the same way.

7.7 σ: $\chi = 3$ (all x and z unshifted, all y become $-y$)

σ': $\chi = 1$ (y_2 and z_2 unshifted, x_2 becomes $-x_2$)

Thus the complete set of characters of the reducible representation is:

C_2	E	C_2	σ	σ'
Γ_1	9	-1	3	1

Because of the basis used, this is termed a Cartesian representation. Reduce this representation using the C_{2v} character table.

7.8 $\Gamma_1 = 3A_1 + A_2 + 3B_1 + 2B_2$

These are the symmetry species of all nine possible molecular movements. From these nine we must now remove the translations and rotations. The translations must belong to A_1, B_1 and B_2 because they must be affected by the group operations in the same way as the x, y, and z directions.

To what species do the three rotations belong?

7.9 A_2, B_1 and B_2. (R_z, R_y and R_x in the character table)

We therefore remove A_1, A_2, $2B_1$ and $2B_2$ from our nine species obtained already and we are left with:

7.10 $2A_1 + B_1$

These are the symmetries of the three vibrational modes of the water molecule (or of any other triatomic C_{2v} molecule).

We can summarise what we have done so far as:

Symmetries of all molecular motions: $3A_1 + A_2 + 3B_1 + 2B_2$

Symmetries of translations $A_1 \qquad + B_1 + B_2$

Symmetries of rotations $A_2 + B_1 + B_2$

∴ Symmetries of vibrations $2A_1 \qquad + B_1$

Do the same analysis for the planar XeF_4 molecule. It belongs to the D_{4h} group and the group operations are given below. What is the reducible representation generated by the set of 15 vectors along the x, y, and z directions for this molecule?

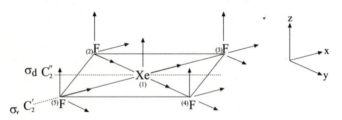

D_{4h}	E	$2C_4$	$C_2(= C_4^2)$	$2C_2'$	$2C_2''$	i	$2S_4$	σ_h	$2\sigma_v$	$2\sigma_d$

7.11

D_{4h}	E	$2C_4$	$C_2(= C_4^2)$	$2C_2'$	$2C_2''$	i	$2S_4$	σ_h	$2\sigma_v$	$2\sigma_d$
Γ_2	15	1	-1	-3	-1	-3	-1	5	3	1

Reduce this representation using the D_{4h} character table. (This may take some time, but it is worthwhile practice.)

7.12 $\Gamma_2 = A_{1g} + A_{2g} + B_{1g} + B_{2g} + E_g + 2A_{2u} + B_{2u} + 3E_u$

What is the total degeneracy of Γ_2, remembering, that A and B are 1-degenerate species, E is 2 degenerate?

7.13   15 i.e. the degeneracy equals the number of vectors in the original basis. This is always true.

Our present 15-degeneracy equals 3×5 for a five-atom molecule. There are, however three translations and three rotations to be removed to leave $3n - 6 = 9$ vibrational modes. What are the symmetry species of the translations?

7.14 $A_{2u} + E_u$ i.e. a singly degenerate translation along z and two equivalent translations along x and y which belong together to the 2-degenerate E_u representation.

What are the symmetry species of the rotations?

7.15 $A_{2g} + E_g$

Take the translations and rotations away from the total Γ_2, and check that the result has a total degeneracy of nine.

$$\Gamma_2 = A_{1g} + A_{2g} + B_{1g} + B_{2g} + E_g + 2A_{2u} + B_{2u} + 3E_u$$

7.16 $\Gamma_2 = A_{1g} + A_{2g} + B_{1g} + B_{2g} + E_g + 2A_{2u} + B_{2u} + 3E_u$

Translations $=$ A_{2u} $+$ E_u

Rotations $=$

            A_{2g} $+ E_g$

$\therefore$ Vibrations $=$

            A_{1g} $+ B_{1g} + B_{2g}$ $+ A_{2u} + B_{2u} + 2E_u$

Total degeneracy $= 9$ for vibrations $(= 3n - 6)$

The irreducible representations we have produced so far represent the symmetries of the nine vibrational modes of the XeF_4 molecule. One of these, for example, is the "breathing" mode in which all four fluorines move out together and then in together. This mode of vibration clearly maintains the full symmetry of the molecule and therefore belongs to the A_{1g} irreducible representation. Other modes of vibration cause distortion of the molecule and are therefore described by other representations.

We now need to determine which, if any, of these modes of vibration are active in the infrared and Raman spectra of the molecule. This is very simple to do if you are prepared to accept a statement of how to do it, rather than to follow a proof. The proof involves calculating the probability of transition in terms of the transition moment integral, and more information on this can be obtained from more advanced textbooks of group theory or spectroscopy.

The rules are simple:

i.   A vibration will be infrared active if it belongs to the same symmetry species as a component of dipole moment, i.e. to the same species as either x, y, or z.

Which of the vibrations of H_2O and of XeF_4 are infrared active?

H_2O vibrations $2A_1 + B_1$
XeF_4 vibrations $A_{1g} + B_{1g} + B_{2g} + A_{2u} + B_{2u} + 2E_u$

7.17 H_2O: all three are active, because z belongs to A_1 and x belongs to B_1.

XeF_4: $A_{2u} + 2E_u$ are active, i.e. in both molecules there should be three i.r. active bands. N.B. $2A_1$ indicates two different vibrations (non degenerate) of the same symmetry. $2E_u$ indicates again two bands, but each one consists of two degenerate vibrations.

The Raman rule is as follows:

ii. A vibration will be Raman active if it belongs to the same symmetry species as a component of polarisability, i.e. to one of the binary products, x^2, y^2, z^2, xy, xz, yz or to a combination of products such as $x^2 - y^2$.

Which vibrations of H_2O and of XeF_4 are Raman active?

7.18 H_2O: all three are active because x^2, y^2, and z^2 belong to A_1 and xz belongs to B_1

XeF_4: A_{1g}, B_{1g}, B_{2g} are Raman active.

We may summarise tnese results as follows:

H_2O: 3 i.r., 3 Raman, 3 coincidences, i.e. the frequency of the i.r. absorptions and of the Raman shifts are identical.

XeF_4?

7.19 XeF_4: 3 i.r., 3 Raman, no coincidences, i.e. the frequencies of
the i.r. absorptions and of the Raman shifts do not coincide
at all.

This is an example of a general effect called the exclusion
rule, Raman shifts and i.r. frequencies never coincide in a
molecule with a centre of symmetry. This occurs because the
x, y, and z directions are always antisymmetric to inversion
through the centre, and belong to representations given a
subscript u, while the binary products are always symmetric
to i and belong to g representations.

Group Theory can also be used to find the nature of the
vibrational mode belonging to each irreducible representation. This topic is dealt with in the next programme.

- -

We will now look at a vibrational analysis of the ammonia
molecule since this illustrates a further feature of the application of Group Theory to molecular vibrations. The 12-arrow
basis for our Cartesian representation is:

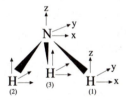

C_{3v} E $2C_3$ $3\sigma_v$

What are the characters of the representation of E and of
one of the planes (chose the xz plane passing through H(1)
and N).

7.20   E: 12 (all arrows are unshifted).
σ: 2 (x and z are unshifted on two atoms, y becomes $-y$).

The C_3 operation clearly shifts all the arrows on the hydrogens, so we only need to consider the arrows on nitrogen.
The z arrow is clearly unaffected and will contribute +1 to
the character. Try to work out the character of the representation of C_3. (Do not take too long if you get stuck — its
rather tricky!)

7.21 C_3: 0

We have already seen that z contributes $+1$ to this, so x and y together must contribute -1. On rotation by a third of a turn (120°), the arrows, looking down the z axis, appear as follows:

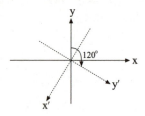

The new y co-ordinate of a point is then dependent on both the old x and the old y co-ordinates, and can be obtained by resolution as:

new $x = x \cos 120° - y \sin 120°$
new $y = x \sin 120° + y \cos 120°$

Remember that z is unshifted by the C_3 operation, and write out the full 3×3 matrix which operates on the matrix $\begin{pmatrix} x \\ y \\ z \end{pmatrix}$.

7.22 $\begin{pmatrix} \cos 120° & -\sin 120° & 0 \\ \sin 120° & \cos 120° & 0 \\ 0 & 0 & 1 \end{pmatrix} \begin{pmatrix} x \\ y \\ z \end{pmatrix} = \begin{pmatrix} x' \\ y' \\ z' \end{pmatrix}$

Since $\cos 120° = -\frac{1}{2}$, this matrix has a character of zero, and the complete set of characters of the Cartesian representation is:

C_{3v}	E	$2C_3$	3σ
Γ_3	12	0	2

Rotation about z through any angle θ can be represented by a matrix

$\begin{pmatrix} \cos\theta & -\sin\theta & 0 \\ \sin\theta & \cos\theta & 0 \\ 0 & 0 & 1 \end{pmatrix}$

but it is rather troublesome to work out the sines and cosines for each individual case. It is easier to consider the atoms in two sets for each symmetry operation:

i. Atoms which are shifted by the operation contribute nothing to the character of the Cartesian representation.

ii. *Each* atom *un*shifted by the operation contributes an amount f(R) to the character of the Cartesian representation where f(R) depends on the operation as follows:

Operation:	E	σ	i	C_2	C_3	C_4	C_5	C_6
f(R) :	3	1	-3	-1	0	1	1.618	2

Operation:	S_3	S_4	S_5	S_6	S_8
f(R) :	-2	-1	0.382	0	0.414

For any C_n, $f(R) = 1 + 2\cos\dfrac{2\pi}{n}$

For any S_n, $f(R) = -1 + 2\cos\dfrac{2\pi}{n}$

This table has been worked out by using similar considerations to those used above for the ammonia molecule.

Use the table to set up the characters of the Cartesian representation of ammonia:

C_{3v}	E	$2C_3$	$3\sigma_v$

7.23

C_{3v}	E	$2C_3$	$3\sigma_v$
Γ_3	12	0	2

E :   4 atoms unshifted,   f(R) = 3, $= 4 \times 3 = 12$
C_3:   1 atom unshifted,   f(R) = 0, $= 1 \times 0 = 0$
σ_v: 2 atoms unshifted,   f(R) = 1, $= 2 \times 1 = 2$

Use the table to set up the characters of the Cartesian representation of CH_4:

T_d	E	$8C_3$	$3C_2$	$6S_4$	$6\sigma_d$

7.24

T_d	E	$8C_3$	$3C_2$	$6S_4$	6σ
Γ_4	15	0	-1	-1	3

If you require further practice at setting up Cartesian representations, you could use the table to set up the representations for water and xenon tetrafluoride discussed earlier.

If you require further practice at finding the number of infrared and Raman bands predicted for a particular molecule, you could confirm that ammonia has four infrared and four coincident Raman bands while methane has two infrared and four Raman bands, two of which are coincident with the infrared bands.

- -

In the final section of this programme we shall look at a particular vibration, such as a carbonyl stretch, occurring in a well defined part of the spectrum, and use Group Theory to predict the number of active bands in this particular region.

The substituted metal carbonyl shown below will undoubtedly absorb in the 1700–2000 cm^{-1} region, the question we wish to answer is, how many bands will there be in the C–O stretching region?

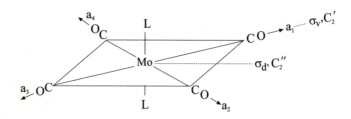

The four-arrow basis shown can be used to represent the carbonyl stretching vibrations. Find the set of characters of the representation obtained by using this basis:

D_{4h} E $2C_4$ $C_2(=C_4^2)$ $2C_2'$ $2C_2''$ i $2S_4$ σ_h $2\sigma_v$ $2\sigma_d$

7.25

D_{4h}	E	$2C_4$	$C_2(=C_4^2)$	$2C_2'$	$2C_2''$	i	$2S_4$	σ_h	$2\sigma_v$	$2\sigma_d$
Γ_5	4	0	0	2	0	0	0	4	2	0

This type of problem is easier than generating the Cartesian representation because the arrows can never be transformed into minus themselves.

Reduce this representation.

7.26 $\Gamma_5 = A_{1g} + B_{1g} + E_u$

Our basis (a_1 to a_4) only included stretching of the C–O bonds, so these three irreducible representations are the representations to which the various C–O stretches belong. We do not in this case need to remove translations or vibrations simply because we did not put them in when setting up the basis of the representation.

Decide, from the character table, how many infrared and Raman active bands there will be in the C–O stretching region.

7.27   1 infrared band (E_u)
2 Raman bands (A_{1g} and B_{1g})

Do the same analysis for the *cis* isomer of the same complex, find how many bands it will have in the C–O region:

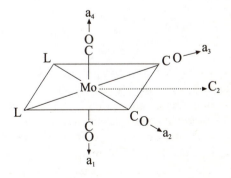

C_{2v} E C_2 σ σ'

7.28   4 infrared bands.
        4 Raman bands (all coincident).

        i.e.

C_{2v}	E	C_2	σ	σ'
Γ_6	4	0	2	2

$\Gamma_6 = 2A_1 + B_1 + B_2$

All these vibrations are active in both infrared and Raman.

Finally, consider the two possible isomers of a metal tri-carbonyl:

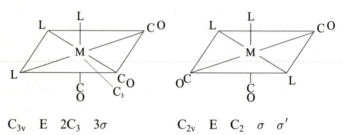

    C_{3v} E $2C_3$ 3σ C_{2v} E C_2 σ σ'

Use the method just developed to find the number of Raman and infrared bands in each isomer.

7.29 C_{3v}: 2 infrared bands $\left.\right\}$ coincident, $A_1 + E$
              2 Raman bands

        C_{2v}: 3 infrared bands $\left.\right\}$ coincident, $2A_1 + B_1$
              3 Raman bands

In general, a set of n CO groups will give rise to n possible C–O stretching modes. The number of observed spectral bands, however, may well be less than n if symmetry makes some modes degenerate or inactive. The use of Group Theory simply formalises this statement and allows precise calculations to be made.

You should now be able to use Group Theory to find the number of infrared and Raman active vibrations in a molecule, and to find the number of active vibrations in a characteristic region of the infrared or Raman spectrum. These topics are the subject of the test which follows.

Applications to Molecular Vibration Test

(You may, if you wish, use the table of f(R) in frame 7.22.)

1. Find the number, and symmetry species, of the Raman and infrared active vibrations of the fumarate ion (C_{2h}):

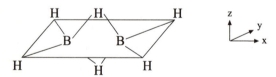

The ion lies in the xy plane. The C_2 axis is the z axis.

2. Find the number, and symmetry species, of the Raman and infrared active vibrations of boron trichloride (D_{3h}):

$$
\begin{array}{c}
Cl \\
| \\
B \\
Cl \quad \quad Cl
\end{array}
$$

3. Find the number of terminal B–H stretching vibrations which are active in the infrared and Raman spectra of diborane (D_{2h}):

Answers

1.   Reducible representation:

C_{2h}	E	C_2	i	σ_h
	30	0	0	10

1 mark

This reduces to: $10A_g + 5B_g + 5A_u + 10B_u$ *1 mark*

Rotations $A_g + 2B_g$

Translations $A_u + 2B_u$

∴ Vibrations $9A_g + 3B_g + 4A_u + 8B_u$ *1 mark*

i.r. active $4A_u + 8B_u$ *1 mark*

Raman active $9A_g + 3B_g$ *1 mark*

2.   Reducible representation:

D_{3h}	E	$2C_3$	$3C_2$	σ_h	$2S_3$	$3\sigma_v$
	12	0	−2	4	−2	2

1 mark

This reduces to: $A_1' + A_2' + 3E' + 2A_2'' + E''$ *1 mark*

Rotations A_2' $+ E''$

Translations $E' + A_2''$

∴ Vibrations A_1' $+ 2E' + A_2''$ *1 mark*

i.r. active $2E' + A_2''$ *1 mark*

Raman active A_1' $+ 2E'$ *1 mark*

3.   Reducible representation:

D_{2h}	E	$C_2(z)$	$C_2(y)$	$C_2(x)$	i	$\sigma(xy)$	$\sigma(xz)$	$\sigma(yz)$
	4	0	0	0	0	4	0	0

1 mark

This reduces to: $A_g + B_{1g} + B_{2u} + B_{3u}$ *1 mark*

i.r. active $B_{2u} + B_{3u}$ *1 mark*

Raman active $A_g + B_{1g}$ *1 mark*

_____

Total 14 marks

Applications to Molecular Vibration

Revision Notes

The application of Group Theory to molecular vibrations can be summarised in three rules:

i.   Use an appropriate *basis* to find a set of characters of a reducible representation of the point group.

ii.   *Reduce* this representation to its constituent irreducible representations.

iii.   *Interpret* the results.

The initial choice of the basis is crucial. In essence this determines the question we are asking the theory to answer. If this is correct the rest of the process follows easily.

A complete vibrational analysis starts with a set of three Cartesian displacement vectors on each atom as the basis. It is then necessary to subtract the irreducible representations to which translations and rotations belong, in order to find the irreducible representations to which the vibrations belong.

If an atom is moved by a symmetry operation, that atom contributes nothing to the character of the resulting reducible representation. If, however, an atom is *un*shifted by a symmetry operation, the contribution of that atom to the character of the reducible representation is given by the quantity $f(R)$. A table of, values of $f(R)$ for various symmetry operations appears in frame 7.22.

The irreducible representations to which specified vibrations (e.g. C–O stretches) belong can be found by taking C–O bond stretching as the basis of the representation. In this case it is not necessary to remove translations or rotations because they are not included in the basis.

Molecular vibrations are:

i.   Infrared active if they belong to the same irreducible representation as x or y or z.

ii.   Raman active if they belong to the same irreducible representation as a binary product such as xy, z^2, $x^2 - y^2$ etc.

Linear Combinations

Objectives

After completing this programme you should be able to:

1. Find the combinations of bond stretching vibrations which form the bond stretching vibrational modes of a molecule.
2. Find the symmetry adapted linear combinations of orbitals suitable for combining with the atomic orbitals of a central atom to form molecular orbitals.
3. Find the form of the wave functions of hybrid orbitals.
4. Normalise any of the above functions.
5. Confirm the orthogonality of normalised functions.

All five objectives are tested at the end of the programme.

Assumed Knowledge

A knowledge of the preceding programmes is assumed.

8.1   Up to now we have not looked in great depth at topics like orbitals or molecular vibrations. We have, for instance, seen that a set of sp^2 hybrids is needed to form a triangular set of bonds but we have not found the wave functions of each orbital. We have found the symmetry properties of various molecular vibrations and have decided on their infrared and Raman activities. We have not, however, worked out the actual form of each vibrational mode. Symmetry theory can help us solve these problems if we extend the mathematics a little using a technique known as the projection operator method.

This approach will be illustrated by a simple example.

Use the stretching of the O–H bonds of water as the basis for a reducible representation of the C_{2v} point group.

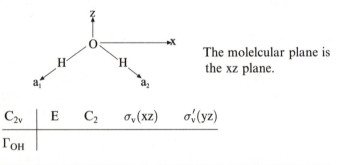

The molelcular plane is the xz plane.

C_{2v}	E	C_2	$\sigma_v(xz)$	$\sigma_v'(yz)$
Γ_{OH}				

8.2

C_{2v}	E	C_2	$\sigma_v(xz)$	$\sigma_v'(yz)$
Γ_{OH}	2	0	2	0

Reduce this representation using the C_{2v} character table.

8.3 $\Gamma_{OH} = A_1 + B_1$

One way in which the O–H bonds can vibrate is for both hydrogens to move out together and both to move in together:

Mode 1: Symmetric stretch.

Suggest another way in which the atoms could vibrate.

8.4

Mode 2: Antisymmetric stretch.

As one atom moves out the other moves in.

Use the character table to find the symmetry species of each of these vibrational modes.

8.5   Symmetric stretch: A_1 (the distorted molecule looks the same after applying any of the group operations).

Antisymmetric stretch: B_1 (the phase of the vibration is reversed by the operations C_2 and $\sigma_v(yz)$ which are represented by -1 in the character table).

We have therefore found the form of the two O–H stretching modes of vibration of the water molecule. Moreover, we can see that the A_1 mode will cause an oscillating dipole moment in the z direction and the B_1 mode will cause an oscillating dipole moment in the x direction. An oscillating dipole moment is a requirement for infrared activity and the directions of these correspond to the directional properties of A_1 and B_1 shown in the character table.

For a simple molecule like water it is a very easy matter to see intuitively the forms of the vibrational modes and to check these from the character table. For more complex molecules, however, we must use a more rigorous approach called the projection operator method. We will look at this in three steps:

Step 1: Draw the set of vectors which formed the basis of the reducible representation.

Step 2: Select one of these vectors as a generating vector and find the result of operating on it by each of the group operations. We will use the arrow a_1 as our generating vector.

We now need the result of applying each of the group operations to a_1. In this case the identity operation leaves a_1 unchanged but the C_2 rotation moves a_1 over to a_2:

C_{2v}	E	C_2	$\sigma_v(xz)$	$\sigma_v'(yz)$
Vector a_1 becomes:	a_1	a_2		

Complete this row.

8.6

C_{2v}	E	C_2	$\sigma_v(xz)$	$\sigma_v'(yz)$
Vector a_1 becomes:	a_1	a_2	a_1	a_2

Step 3: For each irreducible representation we now multiply each of the above results by the character of the irreducible representation in the character table and sum the results.

For example, the form of the A_1 vibration is found by multiplying each result by the character of the A_1 representation in the character table and summing the result:

C_{2v}	E	C_2	$\sigma_v(xz)$	$\sigma_v'(yz)$
	a_1	a_2	a_1	a_2

Character table:

C_{2v}	E	C_2	$\sigma_v(xz)$	$\sigma_v'(yz)$
A_1	1	1	1	1
A_2	1	1	-1	-1
B_1	1	-1	1	-1
B_2	1	-1	-1	1

A_1 vibration: $a_1 \times 1 + a_2 \times 1 + a_1 \times 1 + a_2 \times 1 = 2a_1 + 2a_2$

Similarly:

A_2 vibration: $a_1 \times 1 + a_2 \times 1 + a_1 \times (-1) + a_2(-1) = 0$

Complete this calculation for B_1 and B_2.

8.7

B_1 vibration:

$$a_1 \times 1 + a_2 \times (-1) + a_1 \times 1 + a_2 \times (-1) = 2a_1 - 2a_2$$

B_2 vibration:

$$a_1 \times 1 + a_2 \times (-1) + a_1 \times (-1) + a_2 \times 1 = 0$$

The result: This shows that we have an A_1 and a B_1 vibration but no A_2 or B_2 bond stretching modes, as expected. In general if there is no mode of vibration with a particular symmetry, the result of applying the above procedure will be to produce an answer of zero. This is a useful check on our arithmetic!

You may be concerned that the projection operator procedure has produced modes of vibration described as $(2a_1 + 2a_2)$ rather than just $(a_1 + a_2)$. This matter is easily resolved because all such combinations of bond stretches must be normalised, i.e. the sum of the squares of the coefficients of the vectors must equal 1. We can achieve this by making each coefficient $\frac{1}{\sqrt{2}}$ i.e. the vibrational modes are:

$A_1:$ $\dfrac{1}{\sqrt{2}}(a_1 + a_2)$

$B_1:$ $\dfrac{1}{\sqrt{2}}(a_1 - a_2)$

Try to normalise the following vibrations of an octahedral molecule:

$A_{1g}:$ $(a_1 + a_2 + a_3 + a_4 + a_5 + a_6)$
$E_g:$ $(-a_1 + a_2 - a_3 + a_4)$
$E_g:$ $(-a_1 - a_2 - a_3 - a_4 + 2a_5 + 2a_6)$

8.8 A_{1g}: $\dfrac{1}{\sqrt{6}}\,(a_1 + a_2 + a_3 + a_4 + a_5 + a_6)$

E_g: $\dfrac{1}{2}\,(-a_1 + a_2 - a_3 + a_4)$

A_g: $\dfrac{1}{\sqrt{12}}\,(-a_1 - a_2 - a_3 - a_4 + 2a_5 + 2a_6)$

In each case the sum of the squares of the coefficients equals one, i.e.

A_{1g}: $\dfrac{1}{6} + \dfrac{1}{6} + \dfrac{1}{6} + \dfrac{1}{6} + \dfrac{1}{6} + \dfrac{1}{6}$ $= 1$

E_g: $\dfrac{1}{4} + \dfrac{1}{4} + \dfrac{1}{4} + \dfrac{1}{4}$ $= 1$

E_g: $\dfrac{1}{12} + \dfrac{1}{12} + \dfrac{1}{12} + \dfrac{1}{12} + \dfrac{4}{12} + \dfrac{4}{12} = 1$

The projection operator method can be used to solve other problems beside molecular vibrations. For example, let us set up the molecular orbitals of the water molecule formed by the combination of hydrogen 1s orbitals with the atomic orbitals of oxygen.

We must first find the combination of hydrogen 1s orbitals which transform according to the different symmetry species of the C_{2v} point group. As **Step 1**, therefore, we choose the two hydrogen 1s orbitals as our basis. This will give a representation reducible to $A_1 + B_2$. For **Step 2**, we choose orbital ϕ_1 as our generating vector:

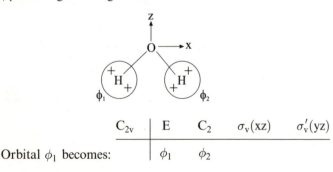

C_{2v}	E	C_2	$\sigma_v(xz)$	$\sigma_v'(yz)$
Orbital ϕ_1 becomes:	ϕ_1	ϕ_2		

Complete this table.

8.9

C_{2v}	E	C_2	$\sigma_v(xz)$	$\sigma_v'(yz)$
Orbital ϕ_1 becomes:	ϕ_1	ϕ_2	ϕ_1	ϕ_2

It should now be apparent that the result is just the same as the O–H bond vibration example worked through earlier and will give as the two combined orbitals:

A_1: $\dfrac{1}{\sqrt{2}}(\phi_1 + \phi_2)$

B_1: $\dfrac{1}{\sqrt{2}}(\phi_1 - \phi_2)$

These will overlap with the oxygen orbitals of the same symmetry as follows:

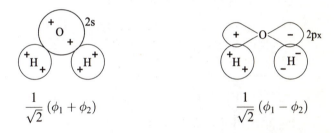

$\dfrac{1}{\sqrt{2}}(\phi_1 + \phi_2)$ $\dfrac{1}{\sqrt{2}}(\phi_1 - \phi_2)$

We will now use the projection operator method on a system where the end result is rather less obvious. We shall try to find the form of the bond stretching vibrations of a flat triangular molecule such as BCl_3 or an ion such as CO_3^{2-} or NO_3^-.

What is the point group of these examples?

8.10 D_{3h} If you are still unsure of this basic idea, have another look at frame 2.22.

If we are interested in the bond stretching vibrations of BCl_3, **Step 1** requires us to use a set of vectors representing the bond stretches as the basis for our reducible representation as we did in frames 7.24 to 7.29.

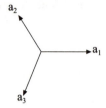

Use these arrows to set up a reducible representation of D_{3h} by completing the following:

	E	$2C_3$	$3C_2$	σ_h	$2S_3$	$3\sigma_v$
Γ_{B-Cl}	3	0				

8.11

	E	$2C_3$	$3C_2$	σ_h	$2S_3$	$3\sigma_v$
Γ_{B-Cl}	3	0	1	3	0	1

Reduce this to its irreducible representations.

8.12 $\Gamma_{B-Cl} = A_1' + E'$

This tells us that B–Cl stretching gives rise to three modes of vibration, one of A_1' symmetry and a degenerate pair of E' symmetry. We now wish to find the form of these vibrational modes.

As **Step 2**, let us use a_1 as the generating vector for the projection operator method. Unfortunately it is not possible to group the symmetry operations into classes; rather we shall have to consider the effect of each of the 12 operations of the group individually. We shall therefore consider the rotations C_3 and S_3 to be anticlockwise and will label the C_2 rotations and vertical planes as $C_2(1)$, $\sigma_v(1)$ etc. to indicate the relevant axis or plane.

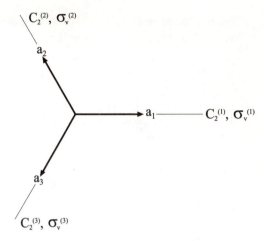

Applying each operation to a_1 we obtain:

D_{3h}	E	C_3	C_3^2	$C_2(1)$	$C_2(2)$	$C_2(3)$	σ_h	S_3	S_3^5	$\sigma_v(1)$	$\sigma_v(2)$	$\sigma_v(3)$
a_1 becomes	a_1	a_2			a_3				a_3		a_3	

Complete this table.

8.13

D_{3h}	E	C_3	C_3^2	$C_2(1)$	$C_2(2)$	$C_2(3)$	σ_h	S_3	S_3^5	$\sigma_v(1)$	$\sigma_v(2)$	$\sigma_v(3)$
a_1 becomes	a_1	a_2	a_3	a_1	a_3	a_2	a_1	a_2	a_3	a_1	a_3	a_2

We can now move to **Step 3** and use the character table to find the form of each mode of vibration generated by this vector. This is quite a long business but the following will give you a start:

D_{3h}	E	C_3	C_3^2	$C_2(1)$	$C_2(2)$	$C_2(3)$	σ_h	S_3	S_3^5	$\sigma_v(1)$	$\sigma_v(2)$	$\sigma_v(3)$
A_1':	$a_1 \times 1$	$a_2 \times 1$	$a_3 \times 1$	$a_1 \times 1$	$a_3 \times 1$							
										Sum =		
A_2'	a_1	a_2	a_3	$-a_1$	$-a_3$	$-a_2$	a_1	a_2	a_3	$-a_1$	$-a_3$	$-a_2$
										Sum = 0		
E':	$2a_1$	$-a_2$	$-a_3$	0	0	0	$2a_1$					
										Sum =		
A_1'':												
A_2'':												
E'':												

Complete this calculation.

8.14   **The result**: the result of this should be that only two symme-
        try species give a non-zero result:

$$A_1' = 4a_1 + 4a_2 + 4a_3$$
$$E' = 4a_1 - 2a_2 - 2a_3$$

Normalise these results i.e. express them in a form such that
the sum of the squares of the coefficients equals unity.

8.15

$$A_1' = \frac{1}{\sqrt{3}}(a_1 + a_2 + a_3)$$

$$E' = \frac{1}{\sqrt{6}}(2a_1 - a_2 - a_3)$$

This gives us the form of two modes of vibration, the totally
symmetric A_1' or "breathing" mode and an E' mode:

A_1' E'

What is the degeneracy of a representation labelled E'?
(HINT: the character of the identity operation will help
here.)

8.16   Two. There is therefore another E' mode of vibration which
        has an identical vibrational frequency but in which the atoms
        move differently. Our problem now is to find this mode.

        The initial temptation is to select another vector (say a_2) as
        the generating vector and repeat the above procedure. Don't
        go all through this but try to write down the result you would
        expect this to give.

8.17   We would expect this to give a similar result rotated through
        one third of a turn, i.e.

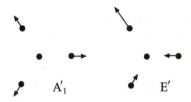

A'_1 E'

This result is not acceptable, partly because we could go on
to generate a third E' mode and it is only 2-degenerate, but
also because the different E' modes would not be orthogonal
to each other. We will discuss orthogonality later but for now
we must introduce another step in the procedure:

Step 3A: if there is a degenerate representation in the group,
select another generating vector at right angles (orthogonal)
to the first and repeat **Steps 2 and 3** for the degenerate representation only.

To do this we shall need to add a few more vectors to our
diagram including one at right angles to a_1 which we shall use
as the second generating vector.

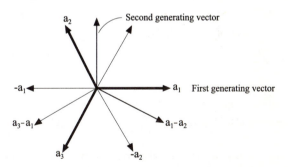

In this diagram, the vector $-a_1$ is pointing in the opposite
direction to a_1. The vector pointing between $-a_1$, and a_3 is
their resultant and can be labelled $(a_3 - a_1)$. Note that this
notation merely describes the direction of the vector. The
length is unimportant at this stage since we shall apply the
normalisation condition to all our results later.

Using this nomenclature, what is the second generating
vector?

8.18   It is $(a_2 - a_3)$ since it lies between a_2 and $-a_3$ and is their resultant.

Use the vector $(a_2 - a_3)$ to apply the projection operator method as you did in frame 8.12 by completing the table: (you may need to add and label additional vectors).

D_{3h}	E	C_3	C_3^2	$C_2(1)$	$C_2(2)$	$C_2(3)$	σ_h	S_3	S_3^5	$\sigma_v(1)$	$\sigma_v(2)$	$\sigma_v(3)$
$a_2 - a_3$ becomes	$+a_2 - a_3$	$+a_1 - a_2$			$-a_3 + a_1$					$-a_2 + a_3$		

8.19

D_{3h}	E	C_3	C_3^2	$C_2(1)$	$C_2(2)$	$C_2(3)$
$a_2 - a_3$ becomes	$+a_2 - a_3$	$+a_3 - a_1$	$+a_1 - a_2$	$-a_2 + a_3$	$-a_1 + a_2$	$-a_3 + a_1$

	σ_h	S_3	S_3^5	$\sigma_v(1)$	$\sigma_v(2)$	$\sigma_v(3)$
	$+a_2 - a_3$	$+a_3 - a_1$	$+a_1 - a_2$	$-a_2 + a_3$	$-a_1 + a_2$	$-a_3 + a_1$

We now only need to multiply this by the characters in the E' representation in the character table to obtain the form of the second E' mode of vibration. This gives us:

D_{3h}	E	C_3	C_3^2	$C_2(1)$	$C_2(2)$	$C_2(3)$	σ_h	S_3	S_3^5	$\sigma_v(1)$	$\sigma_v(2)$	$\sigma_v(3)$
	$2a_2 - 2a_3$	$-a_1 + a_2$	0					$-a_3 + a_1$			0	

Complete this table

8.20

D_{3h}	E	C_3	C_3^2	$C_2(1)$	$C_2(2)$	$C_2(3)$	σ_h	S_3	S_3^5	$\sigma_v(1)$	$\sigma_v(2)$	$\sigma_v(3)$
	$2a_2 - 2a_3$	$-a_3 + a_1$	$-a_1 + a_2$	0	0	0	$+2a_2 - 2a_3$	$-a_3 + a_1$	$-a_1 + a_2$	0	0	0

Now add up the result to find how many a_1 how many a_2 and how many a_3 describe the mode of vibration.

8.21   **The result:** the sum is $(6a_2 - 6a_3)$

Normalise this result.

8.22   The second E' mode is:

$$\frac{1}{\sqrt{2}}(a_2 - a_3) \text{ i.e.}$$

Our complete set of bond stretching vibrations is therefore:

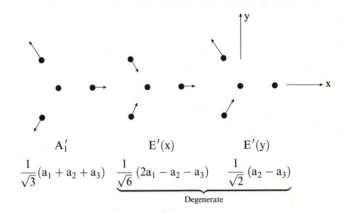

$$\begin{array}{ccc}
A_1' & E'(x) & E'(y) \\
\frac{1}{\sqrt{3}}(a_1 + a_2 + a_3) & \frac{1}{\sqrt{6}}(2a_1 - a_2 - a_3) & \frac{1}{\sqrt{2}}(a_2 - a_3)
\end{array}$$

Degenerate

The character table for D_{3h} shows that the A_1' mode is not infrared active but that the E' modes are since they belong to the same representation as the x and y directions. Using the axes shown above we can see that the first E' mode of vibration will give rise to an oscillating dipole moment in the x direction. Bond 1 is oscillating purely along the x axis; bonds 2 and 3 have components along y which cancel out but a small component along x. The second E' mode gives rise to an oscillating dipole moment along the y direction since the x components of bonds 2 and 3 cancel out.

The vibrational modes shown above are, of course, only schematic. When a real molecule or ion vibrates, its centre of gravity remains in the same place, so the central atom must move as well as the outer ones. The extent of this movement will depend on the relative masses of the atoms concerned. Thus the NO_3^- ion, containing atoms of similar masses, will behave differently from BCl_3 where the central atom is very light. The symmetries and general form of the vibrational modes are, however, the same in both cases. Readers wishing to explore the topic of molecular vibrations in more detail should refer to the excellent books by Woodward and by Wilson, Decius and Cross given in the bibliography.

As a final check on the form of our vibrational modes we must ensure that two conditions are satisfied. One is that we have overall used each bond equally and the other is that the functions describing the modes of vibration are orthogonal to each other. These two checks are easily done if we write out the matrix of coefficients of a_1, a_2 and a_3:

	a_1	a_2	a_3
A_1'	$\dfrac{1}{\sqrt{3}}$	$\dfrac{1}{\sqrt{3}}$	$\dfrac{1}{\sqrt{3}}$
$E'(x)$	$\dfrac{2}{\sqrt{6}}$	$\dfrac{-1}{\sqrt{6}}$	$\dfrac{-1}{\sqrt{6}}$
$E'(y)$		$\dfrac{1}{\sqrt{2}}$	$\dfrac{-1}{\sqrt{2}}$

If we add the squares of the coefficients of a_1 we obtain:

$$\frac{1}{3} + \frac{4}{6} = 1$$

Add up the squares of the coefficients of a_2 and a_3 in the same way.

8.23

a_2: $\quad \dfrac{1}{3} + \dfrac{1}{6} + \dfrac{1}{2} = 1$ $\qquad\qquad$ a_3: $\quad \dfrac{1}{3} + \dfrac{1}{6} + \dfrac{1}{2} = 1$

This shows that all three bonds contribute equally to the total picture of the bond stretching vibrations of the molecule as they must since they are all equivalent.

The second condition is that the vibrational modes are orthogonal to each other. To check this we select two modes (say A_1' and $E'(x)$), multiply together the coefficients of each vector and sum the result:

	a_1	a_2	a_3	
$A_1' \times E'(x)$	$\dfrac{1}{\sqrt{3}} \times \dfrac{2}{\sqrt{6}}$	$\dfrac{1}{\sqrt{3}} \times \dfrac{-1}{\sqrt{6}}$	$\dfrac{1}{\sqrt{3}} \times \dfrac{-1}{\sqrt{6}}$	Sum $= 0$

The result should be zero for any pair of vibrational modes.

Check the orthogonality of the other two pairs of modes in the same way:

	a_1	a_2	a_3	
$A_1' \times E'(y)$	$\dfrac{1}{\sqrt{3}} \times 0$	$\dfrac{1}{\sqrt{3}} \times \dfrac{1}{\sqrt{2}}$		Sum =
$E'(x) \times E'(y)$				Sum =

8.24

	a_1	a_2	a_3	
$A_1' \times E'(y)$	$\dfrac{1}{\sqrt{3}} \times 0$	$\dfrac{1}{\sqrt{3}} \times \dfrac{1}{\sqrt{2}}$	$\dfrac{1}{\sqrt{3}} \times \dfrac{-1}{\sqrt{2}}$	Sum = 0
$E'(x) \times E'(y)$	$\dfrac{2}{\sqrt{6}} \times 0$	$\dfrac{-1}{\sqrt{6}} \times \dfrac{1}{\sqrt{2}}$	$\dfrac{-1}{\sqrt{6}} \times \dfrac{-1}{\sqrt{2}}$	Sum = 0

So our results are mutually orthogonal.

In frames 8.12 to 8.22 we used the three vectors a_1, a_2 and a_3 to represent bond stretches and found the combination which had the correct symmetry to describe the bond stretching vibrations of BCl_3. It should be clear that a set of chlorine p-orbitals has the same directional properties as our set of three vectors:

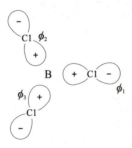

So, without working through the arithmetic again, we can make up the same linear combination of chlorine p-orbitals to combine with the orbitals of boron and form σ-molecular orbitals.

Which boron orbital would interact with the totally symmetric combination of chlorine orbitals?

$$A_1' = \frac{1}{\sqrt{3}}(\phi_1 + \phi_2 + \phi_3)$$

8.25   The spherically symmetrical 2s orbital:

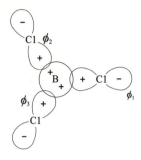

Draw the corresponding diagram for the $E'(x)$ combination interacting with the $2p_x$ orbital of boron.

$$E'(x) = \frac{1}{\sqrt{6}} (2\phi_1 - \phi_2 - \phi_3)$$

8.26

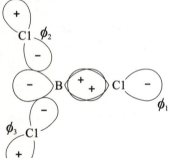

The orbital on $Cl_{(1)}$ has been made larger because its coefficient in the wave function is twice as large as the others.

Draw the corresponding diagram for the other combination,

$$E'(y) = \frac{1}{\sqrt{2}} (\phi_2 - \phi_3)$$

8.27

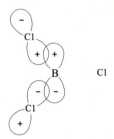

A further use of our set of linear combinations is to regard them as hybrid orbitals of the central boron atom made up of a mixture of 2s, $2p_x$ and $2p_y$ atomic orbitals. The 2s orbital has A_1' symmetry and the character table shows us that the 2p (x and y) orbitals have E' symmetry. We can therefore write:

$$s = \frac{1}{\sqrt{3}} (a_1 + a_2 + a_3) \qquad \ldots 8.1$$

$$p_x = \frac{1}{\sqrt{6}} (2a_1 - a_2 - a_3) \qquad \ldots 8.2$$

$$p_y = \frac{1}{\sqrt{2}} (a_2 - a_3) \qquad \ldots 8.3$$

Where s, p_x and p_y represent the wave functions of the atomic orbitals.

Rearranging equations 8.1 and 8.2 we obtain:

$$\sqrt{3}\,s = a_1 + a_2 + a_3$$

$$\sqrt{6}\,p_x = 2a_1 - a_2 - a_3$$

Add together these equations and find a value for a_1.

8.28

$$\sqrt{3}\,s + \sqrt{6}\,p_x = 3a_1$$

$$a_1 = \frac{1}{\sqrt{3}}\,s + \frac{\sqrt{2}}{\sqrt{3}}\,p_x = \frac{1}{\sqrt{3}}\,(s + \sqrt{2}\,p_x)$$

This is the form of the wave function of the sp^2 hybrid orbital of boron pointing along a_1. The equations can be solved to find the other hybrid orbitals as follows:

$$a_2 = \frac{1}{\sqrt{6}}\,(\sqrt{2}\,s - p_x + \sqrt{3}\,p_y)$$

$$a_3 = \frac{1}{\sqrt{6}}\,(\sqrt{2}\,s - p_x - \sqrt{3}\,p_y)$$

Complete the matrix of coefficients of s, p_x and p_y to start the two checks described in frames 8.22 and 8.23.

	s	p_x	p_y
a_1	$\dfrac{1}{\sqrt{3}}$	$\dfrac{\sqrt{2}}{\sqrt{3}}$	0
a_2		$\dfrac{-1}{\sqrt{6}}$	
a_3			

8.29

	s	p_x	p_y
a_1	$\dfrac{1}{\sqrt{3}}$	$\dfrac{\sqrt{2}}{\sqrt{3}}$	0
a_2	$\dfrac{1}{\sqrt{3}}$	$\dfrac{-1}{\sqrt{6}}$	$\dfrac{\sqrt{3}}{\sqrt{6}}$
a_3	$\dfrac{1}{\sqrt{3}}$	$\dfrac{-1}{\sqrt{6}}$	$\dfrac{-\sqrt{3}}{\sqrt{6}}$

Sum the squares of the coefficients of each orbital to demonstrate that each orbital contributes equally to the set of hybrids, i.e.

$$s = \frac{1}{3} + \frac{1}{3} + \frac{1}{3} = 1$$

8.30

$$p_x = \frac{2}{3} + \frac{1}{6} + \frac{1}{6} = 1$$

$$p_y = 0 + \frac{3}{6} + \frac{3}{6} = 1$$

Now check that all pairs of functions are orthogonal by multiplying the corresponding coefficients and showing that they sum to zero, e.g.

$$a_1 \times a_2 = \frac{1}{\sqrt{3}} \times \frac{1}{\sqrt{3}} + \frac{\sqrt{2}}{\sqrt{3}} \times \frac{-1}{\sqrt{6}} + 0 = 0$$

8.31   The other products are:

$$a_1 \times a_3 = \frac{1}{\sqrt{3}} \times \frac{1}{\sqrt{3}} + \frac{\sqrt{2}}{\sqrt{3}} \times \frac{-1}{\sqrt{6}} + 0 = 0$$

$$a_2 \times a_3 = \frac{1}{\sqrt{3}} \times \frac{1}{\sqrt{3}} + \frac{-1}{\sqrt{6}} \times \frac{-1}{\sqrt{6}} + \frac{\sqrt{3}}{\sqrt{6}} \times \frac{-\sqrt{3}}{\sqrt{6}} = 0$$

The three hybrid orbitals, a_1 a_2 and a_3 are therefore orthogonal and the three atomic orbitals, 2s, $2p_x$ and $2p_y$ contribute equally to them.

A Simplified Procedure

In frames 8.10 to 8.12 we used three vectors pointing towards the corners of a triangular molecule as a basis for a reducible representation of the group D_{3h}. This can be used to find the symmetries of the bond stretching vibrations of a molecule like BCl_3 or to demonstrate that a set of sp^2 hybrids is triangular. Most of the programme, however, has been used to show how the projection operator method can be used to find linear combinations of functions. These linear combinations then give us the explicit forms of vibrational modes, hybrid orbitals or molecular orbitals. The projection operator method was just about manageable with the twelve operations of the D_{3h} group but becomes unwieldy with groups containing more operations. In the next section, therefore, we shall try to work intuitively from the character table to find the correct combinations without working through the full arithmetic. We shall stay with the D_{3h} group for this as we have already seen the results for this group.

We start from the fact that the set of three vectors, a_1, a_2 and a_3 is the basis for a representation which reduces to $A_1' + E'$ (frame 8.12).

We now ask ourselves what combination of a_1, a_2 and a_3, could have A_1' symmetry? The character table shows that A_1' is the totally symmetric representation. What combination of a_1, a_2 and a_3 maintains the full symmetry of the molecule?

8.32   Any combination in which they are equally represented, i.e. $a_1 + a_2 + a_3$.

Write this in normalised form.

8.33

$$\frac{1}{\sqrt{3}} (a_1 + a_2 + a_3)$$

We now look at the character table for D_{3h} and note that the E' representation is the one to which both x and y belong.

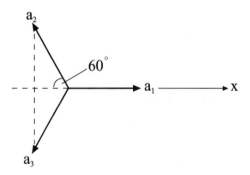

The contribution of each vector along the x direction is simply the projection of that vector onto the x axis which is:

for a_1: $= a_1$

for a_2: $= -a_2 \cos 60° = \dfrac{-1}{2} a_2$

for a_3: $= -a_3 \cos 60° = \dfrac{-1}{2} a_3$

Add these three components together and normalise the result.

8.34 $a_1 - \dfrac{1}{2} a_2 - \dfrac{1}{2} a_3$ or $\dfrac{1}{\sqrt{6}} (2a_1 - a_2 - a_3)$

i.e. the same result as $E'(x)$ obtained previously.

Now let us look at the projection of the vectors onto the y-axis. This is easier because a_1 makes no contribution at all.

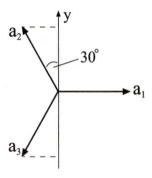

Write down the projection of a_2 and a_3 along the y-axis and normalise the result.

8.35 $a_2 \cos 30° - a_3 \cos 30°$ or, in normalised form, $\dfrac{1}{\sqrt{2}} (a_2 - a_3)$ as obtained previously for $E'(y)$.

This approach obviously has much to commend it especially for groups such as D_{4h} with 16 operations, T_d with 24 or O_h with 48.

Let us look at the application of this less formal method to a tetrahedral set of vectors. The problem is greatly eased by a careful choice of co-ordinate system which relies on the fact that we can construct a tetrahedron by putting one corner at every other corner of a cube.

Use the four vectors a_1, a_2 a_3 and a_4 as a basis for a representation of the T_d group. These vectors could represent the C—H bonds of a methane molecule.

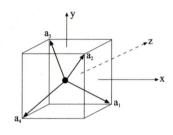

T_d	E	$8C_3$	$3C_2$	$6S_4$	$6\sigma_d$
Γ_{C-H}	4	1			

8.36

T_d	E	$8C_3$	$3C_2$	$6S_4$	$6\sigma_d$
Γ_{C-H}	4	1	0	0	2

Reduce this representation.

8.37 $\Gamma_{C-H} = A_1 + T_2$

If you did not obtain this result, look back at frames 5.26 to 5.30 which go over the process in detail.

We now want to find the combinations of a_1, a_2, a_3 and a_4 which will make our A_1 and T_2 representations. The A_1 case should by now be trivial. Write down the normalised combination.

8.38

A_1: $\quad \dfrac{1}{2}(a_1 + a_2 + a_3 + a_4)$

The other cases are also quite straightforward because the three degenerate T_2 combinations represent the x, y and z directions and the projection of each vector onto the x, y or z axis is simply half the length of side of the cube in each case. We can therefore take this as our unit of length and only consider whether the projection is along the +ve or −ve direction of the axis. Thus the combination corresponding to $T_2(x)$ is:

$T_2(x)$: $\quad \dfrac{1}{2}(a_1 + a_2 - a_3 - a_4)$

Write down the combinations corresponding to $T_2(y)$ and $T_2(z)$.

8.39

$T_2(y)$: $\quad \dfrac{1}{2}(-a_1 + a_2 + a_3 - a_4)$

$T_2(z)$: $\quad \dfrac{1}{2}(a_1 - a_2 + a_3 - a_4)$

We can now write out the matrix of coefficients as:

	a_1	a_2	a_3	a_4
A_1	$\dfrac{1}{2}$	$\dfrac{1}{2}$	$\dfrac{1}{2}$	$\dfrac{1}{2}$
$T_2(x)$	$\dfrac{1}{2}$	$\dfrac{1}{2}$	$\dfrac{-1}{2}$	$\dfrac{-1}{2}$
$T_2(y)$	$\dfrac{-1}{2}$	$\dfrac{1}{2}$	$\dfrac{1}{2}$	$\dfrac{-1}{2}$
$T_2(z)$	$\dfrac{1}{2}$	$\dfrac{-1}{2}$	$\dfrac{1}{2}$	$\dfrac{-1}{2}$

from which it is easy to demonstrate that all four combinations are orthogonal to all others.

We can then go on to use this result to find the form of the bond stretching vibrations of the CH_4 molecule, to construct combinations of hydrogen orbitals or to find the wave functions of the sp^3 hybrid orbitals of carbon. This exercise is left to the reader but the answers are given at the end of the programme.

We will finally apply the less formal method to a square planar system belonging to the D_{4h} point group. We will choose axes such that the four vectors point along the co-ordinate axes.

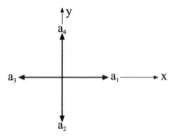

We saw in frames 7.24 to 7.26 that this set of vectors formed the basis of a representation of D_{4h} which could be reduced to $A_{1g} + B_{1g} + E_u$. The A_{1g} (totally symmetric) combination is again $\frac{1}{2}(a_1 + a_2 + a_3 + a_4)$. The character table tells us that the B_{1g} combination has the same symmetry as $x^2 - y^2$ which must mean that we make a_1 and a_3, which point along the x axis, positive but a_2 and a_4, pointing along the y axis, negative:

$$B_{1g} : \frac{1}{2}(a_1 - a_2 + a_3 - a_4)$$

Try to work out the combinations of a_1, a_2, a_3 and a_4 which have $E_u(x)$ and $E_u(y)$ symmetry.

8.40

$$E_u(x): \frac{1}{\sqrt{2}}(a_1 - a_3) \qquad E_u(y): \frac{1}{\sqrt{2}}(a_4 - a_2)$$

The matrix of coefficients is then:

	a_1	a_2	a_3	a_4
A_{1g}	$\frac{1}{2}$	$\frac{1}{2}$	$\frac{1}{2}$	$\frac{1}{2}$
B_{1g}	$\frac{1}{2}$	$\frac{-1}{2}$	$\frac{1}{2}$	$\frac{-1}{2}$
$E_u(x)$	$\frac{1}{\sqrt{2}}$	0	$\frac{-1}{\sqrt{2}}$	0
$E_u(y)$	0	$\frac{-1}{\sqrt{2}}$	0	$\frac{1}{\sqrt{2}}$

from which the orthogonality condition can be seen easily.

Draw the forms of the infrared active $E_u(x)$ and $E_u(y)$ modes of vibration of a square planar molecule by using the coefficients of a_1, a_2, a_3 and a_4 in the above table. Remember that a negative sign means that the direction of the movement of an atom is the reverse of the direction of the original vector.

8.41

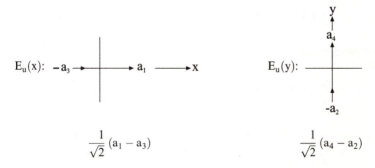

Both modes of vibration cause obvious oscillating dipole moments in the direction of the axis.

We can, however, reasonably ask why the following modes of vibration do not emerge from this analysis:

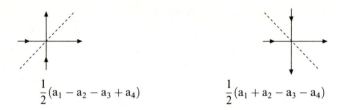

$$\frac{1}{2}(a_1 - a_2 - a_3 + a_4)$$

$$\frac{1}{2}(a_1 + a_2 - a_3 - a_4)$$

Both give obvious oscillating dipoles in the directions shown. The answer is that these combinations would have emerged if we had chosen axes at 45° to the ones used. Alternatively, they can be derived as normalised linear combinations of $E_u(x)$ and $E_u(y)$:

$$E_u(x) + E_u(y) = \frac{1}{2}(a_1 - a_2 - a_3 + a_4)$$

$$E_u(x) - E_u(y) = \frac{1}{2}(a_1 + a_2 - a_3 - a_4)$$

Such linear combinations are perfectly acceptable solutions to our problem but only two orthogonal results may be used.

You should now be able to use the projection operator method to find the form of molecular vibrations, to set up linear combinations of orbitals suitable for the formation of molecular orbitals or to find the form of the wave functions of hybrid orbitals. These topics are the subject of the test which follows.

Conclusion

There is much more to Group Theory than can be covered in a simple introductory text such as this. The subject can also be used to give further insights into the aspects of chemistry considered. You should, however, now be able to tackle some of the more advanced books listed in the bibliography and make reasonably rapid progress with them.

Linear Combinations Test

1.   a. Find the symmetry adapted linear combinations of the
         C—H bond stretches of ethene (D_{2h}) and hence find the
         modes of vibration of the C—H bonds. Use the following
         co-ordinate system:

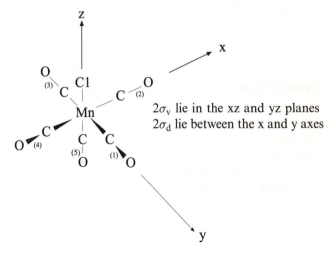

      b. Normalise your result
      c. Show that the different modes are all orthogonal to each
         other.

2.   a. Show that $Mn(CO)_5Cl$ (C_{4v}) has $2A_1 + B_1 + E$ C—O
         stretching modes.
      b. Find the normalised symmetry adapted linear combina-
         tions of the C—O stretches which have these symmetry
         properties. For the two A_1 modes, find one involving
         bond 5 only and one involving bonds 1, 2, 3, and 4 only.
      c. Find linear combinations of the two A_1 modes.
      d. Show that your results are mutually orthogonal.

$2\sigma_v$ lie in the xz and yz planes
$2\sigma_d$ lie between the x and y axes

Answers

1.   a. and b.

D_{2h}	E	$C_{2(z)}$	$C_{2(y)}$	$C_{2(x)}$	i	$\sigma_{(xy)}$	$\sigma_{(xz)}$	$\sigma_{(yz)}$
Γ_{CH}	4	0	0	0	0	0	0	4

$$= A_g + B_{3g} + B_{1u} + B_{2u}$$

1 mark

Normalised linear combinations:

A_g: totally symmetric 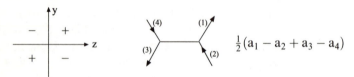 $\frac{1}{2}(a_1 + a_2 + a_3 + a_4)$

1 mark

B_{3g}: same symmetry properties as yz. This function is +ve when y and z are both +ve or both −ve, but −ve if either y or z is −ve:

$\frac{1}{2}(a_1 - a_2 + a_3 - a_4)$

1 mark

B_{1u}: same symmetry properties as z. Hence net displacements along the z direction:

$\frac{1}{2}(a_1 + a_2 - a_3 - a_4)$

1 mark

B_{2u}: same symmetry properties as y. Hence net displacements along the y direction:

$\frac{1}{2}(a_1 - a_2 - a_3 + a_4)$

1 mark

c. Matrix of coefficients:

	a_1	a_2	a_3	a_4
A_g	$\dfrac{1}{2}$	$\dfrac{1}{2}$	$\dfrac{1}{2}$	$\dfrac{1}{2}$
B_{3g}	$\dfrac{1}{2}$	$-\dfrac{1}{2}$	$\dfrac{1}{2}$	$-\dfrac{1}{2}$
B_{1u}	$\dfrac{1}{2}$	$\dfrac{1}{2}$	$-\dfrac{1}{2}$	$-\dfrac{1}{2}$
B_{2u}	$\dfrac{1}{2}$	$-\dfrac{1}{2}$	$-\dfrac{1}{2}$	$\dfrac{1}{2}$

Orthogonality test:

$$A_g \times B_{3g} = \frac{1}{4} - \frac{1}{4} + \frac{1}{4} - \frac{1}{4} = 0$$

$$A_g \times B_{1u} = \frac{1}{4} + \frac{1}{4} - \frac{1}{4} - \frac{1}{4} = 0 \qquad \text{etc.} \qquad \textit{2 marks}$$

2.   a.

C_{4v}	E	$2C_4$	C_2	$2\sigma_v$	$2\sigma_d$
Γ_{CO}	5	1	1	3	1

$$= 2A_1 + B_1 + E \qquad \textit{1 mark}$$

b. First A_1 mode (Bond 5 only): a_5
Second A_1 mode
(Bonds 1, 2, 3 and 4) $\frac{1}{2}(a_1 + a_2 + a_3 + a_4)$

1 mark

B_1 mode (same symmetry properties as $(x^2 - y^2)$ i.e. +ve displacement along both directions of the x axis and −ve displacement along both directions of the y axis):

$\frac{1}{2}(-a_1 + a_2 - a_3 + a_4)$ *1 mark*

E modes (same symmetry properties as x and y, i.e. displacements along the x or y directions):

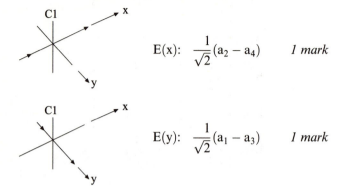

E(x): $\dfrac{1}{\sqrt{2}}(a_2 - a_4)$ *1 mark*

E(y): $\dfrac{1}{\sqrt{2}}(a_1 - a_3)$ *1 mark*

c. Linear combinations of the A_1 modes are:

First + Second:

$$\frac{1}{\sqrt{8}}\,(a_1 + a_2 + a_3 + a_4 + 2a_5)$$ *1 mark*

First − Second:

$$\frac{1}{\sqrt{8}}\,(a_1 + a_2 + a_3 + a_4 - 2a_5)$$ *1 mark*

d. Orthogonality can be shown from either of the matrices of coefficients:

	a_1	a_2	a_3	a_4	a_5
A_1					1
A_1	$\dfrac{1}{2}$	$\dfrac{1}{2}$	$\dfrac{1}{2}$	$\dfrac{1}{2}$	
B_1	$-\dfrac{1}{2}$	$\dfrac{1}{2}$	$-\dfrac{1}{2}$	$\dfrac{1}{2}$	
E(x)		$\dfrac{1}{\sqrt{2}}$		$-\dfrac{1}{\sqrt{2}}$	
E(y)	$\dfrac{1}{\sqrt{2}}$		$-\dfrac{1}{\sqrt{2}}$		

	a_1	a_2	a_3	a_4	a_5
A_1	$\dfrac{1}{\sqrt{8}}$	$\dfrac{1}{\sqrt{8}}$	$\dfrac{1}{\sqrt{8}}$	$\dfrac{1}{\sqrt{8}}$	$\dfrac{2}{\sqrt{8}}$
A_1	$\dfrac{1}{\sqrt{8}}$	$\dfrac{1}{\sqrt{8}}$	$\dfrac{1}{\sqrt{8}}$	$\dfrac{1}{\sqrt{8}}$	$-\dfrac{2}{\sqrt{8}}$
B_1	$-\dfrac{1}{2}$	$\dfrac{1}{2}$	$-\dfrac{1}{2}$	$\dfrac{1}{2}$	
$E(x)$		$\dfrac{1}{\sqrt{2}}$		$-\dfrac{1}{\sqrt{2}}$	
$E(y)$	$\dfrac{1}{\sqrt{2}}$		$-\dfrac{1}{\sqrt{2}}$		

1 mark

Total   *15 marks*

A score of about 10 or more shows a reasonable understanding of the subject but ultimately you should try to get completely correct answers to problems.

Look back now at frames 8.36 to 8.39 in which we started and try to find:

a. the form of the C—H stretching vibrational modes of methane;
b. the linear combinations of hydrogen 1s orbitals which are needed to form molecular orbitals of methane;
c. the wave functions of the sp^3 hybrid orbitals of carbon.

The answers to all these problems are closely related and are given in the following pages.

Results of the Tetrahedral Case

In frame 8.39 we saw that the symmetry adapted combination of four bond vectors in a tetrahedral molecule such as methane gave the matrix of coefficients:

	a_1	a_2	a_3	a_4
A_1	$\frac{1}{2}$	$\frac{1}{2}$	$\frac{1}{2}$	$\frac{1}{2}$
$T_2(x)$	$\frac{1}{2}$	$\frac{1}{2}$	$-\frac{1}{2}$	$-\frac{1}{2}$
$T_2(y)$	$-\frac{1}{2}$	$\frac{1}{2}$	$\frac{1}{2}$	$-\frac{1}{2}$
$T_2(z)$	$\frac{1}{2}$	$-\frac{1}{2}$	$\frac{1}{2}$	$-\frac{1}{2}$

The forms of the bond stretching vibrations are therefore:
A_1 (totally symmetric or 'breathing' mode):

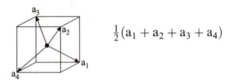

$\frac{1}{2}(a_1 + a_2 + a_3 + a_4)$

$T_2(x)$:
(oscillating dipole moment along the x direction):

$\frac{1}{2}(a_1 + a_2 - a_3 - a_4)$

$T_2(y)$:

$\frac{1}{2}(-a_1 + a_2 + a_3 - a_4)$

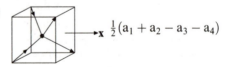

$T_2(z)$:

$\frac{1}{2}(a_1 - a_2 + a_3 - a_4)$

In each of these modes, two atoms move in and two move out. If we add together the three T_2 combinations and normalise the result, we obtain a linear combination:

$$T_2(x, y, z): \qquad \frac{1}{\sqrt{12}}(a_1 + a_2 + a_3 - 3a_4)$$

This represents three atoms moving out while the fourth moves in. The resulting oscillating dipole moment lies in the direction of one of the bonds. This result is a perfectly valid vibrational mode of the molecule but the two other T_2 modes whose oscillating dipoles are at right angles (orthogonal) to it are more complex than the set described above.

The combination of hydrogen 1s orbitals that will form molecular orbitals with the central carbon atom can also be obtained from the matrix of coefficients. The A_1 combination will interact with the spherical 2s orbital while the other combinations will interact with the 2p orbitals, e.g.

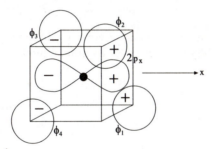

$$T_2(x) = \frac{1}{2}(\phi_1 + \phi_2 - \phi_3 - \phi_4)$$

The wave functions of the four sp^3 hybrid orbitals of carbon can also be obtained by rearranging the equations to find values for a_1, a_2, a_3 and a_4, giving the result:

$$\phi_1 = \frac{1}{2}(s + p_x - p_y + p_z)$$

$$\phi_2 = \frac{1}{2}(s + p_x + p_y - p_z)$$

$$\phi_3 = \frac{1}{2}(s - p_x + p_y + p_z)$$

$$\phi_4 = \frac{1}{2}(s - p_x - p_y - p_z)$$

Linear Combinations

Revision Notes

The explicit form of molecular vibrations, hybrid orbitals or the combinations of atomic orbitals suitable for forming molecular orbitals can be found by applying the projection operator method. The steps in this method are as follows:

Step 1: Choose a set of vectors forming the basis of the reducible representation.

Step 2: Select one of these as a generating vector and find the result of operating on it by each of the group operations.

Step 3: For each symmetry species:
Multiply each of the above results by the character of the irreducible representation in the character table and sum the result.

Step 3A: If there is a degenerate representation in the group, select another orthogonal vector and repeat steps 2 and 3 for the degenerate representation only.

The result: The resulting linear combination of vectors gives the form of each irreducible representation.

Sets of linear combinations must be normalised and orthogonal.

This process can become very long winded for groups with many symmetry operations, so it can often be shortened as follows:

1. Choose a set of vectors to represent the problem to be solved, use these as a basis of a reducible representation of the group and reduce this to its irreducible representations (i.e. the normal procedure covered in Programmes 6 and 7).

2. Using the character table, write down combinations of the basis vectors having the same symmetry properties as the irreducible representations found above. If necessary, use the projection of the basis vector along a specified direction.

3. Normalise the resulting combinations and check for orthogonality.

Bibliography

P W Atkins, M S Child and C S G Phillips, Tables for Group Theory,
Oxford University Press, 1970

F A Cotton, Chemical Applications of Group Theory, (2nd Ed),
Wiley Interscience, 1971

G Davidson, Introductory Group Theory for Chemistry,
Elsevier, 1971

G Davidson, Group Theory for Chemists,
Macmillan, 1991

J D Donaldson and S D Ross, Symmetry and Stereochemistry,
Intertext, 1972

S F A Kettle, Symmetry and Structure,
Wiley, 1985

J A Salthouse and M J Ware, Point Group Character Tables and Related Data,
Cambridge University Press, 1972

D S Urch, Orbitals and Symmetry,
Penguin, 1970

R McWeeny, Symmetry, an Introduction to Group Theory,
Pergamon, 1963

E P Wigner, Group Theory,
Academic Press, 1959

E B Wilson, Jr, J C Decius and P C Cross, Molecular Vibrations,
McGraw-Hill, 1955

L A Woodward, Introduction to the Theory of Molecular Vibrations and
Vibrational Spectroscopy,
Oxford University Press, 1972

Mathematical Data for use with Character Tables

1. **Character Tables containing Complex Numbers**

   In some character tables the two-degenerate, E representation consists of two lines of numbers, some of which are complex e.g.:

C_3	E	C_3	C_3^2
A	1	1	1
E	$\begin{cases} 1 \\ 1 \end{cases}$	$\begin{array}{c} \exp(2\pi i/3) \\ \exp(-2\pi i/3) \end{array}$	$\begin{array}{c} \exp(-2\pi i/3) \\ \exp(2\pi i/3) \end{array}$

   This is done so that the characters do, in fact, satisfy various theorems of group theory. In practical use, however, the two lines are added up and the following relationships will be found helpful:

   $$\varepsilon = \exp(2\pi i/n) = \cos(2\pi/n) + i\sin(2\pi/n)$$
   $$\varepsilon^* = \exp(-2\pi i/n) = \cos(2\pi/n) - i\sin(2\pi/n)$$

   Hence: $\exp(2\pi i/n) + \exp(-2\pi i/n) = 2\cos(2\pi/n)$

   The table can therefore be used as if it read:

C_3	E	C_3	C_3^2
A	1	1	1
E	2	$2\cos(2\pi/3)$	$2\cos(2\pi/3)$

   i.e.

C_3	E	C_3	C_3^2
A	1	1	1
E	2	-1	-1

2.   Character Tables for Groups containing a C_5 Axis

Groups containing a five fold axis have character tables containing cos 72° ($2\pi/5$) and cos 144° ($4\pi/5$) or exponentials which adds up to give these quantities. The following relationships will avoid the necessity of working with cumbersome decimal numbers:

$$2\cos 72° = \tau - 1$$
$$2\cos 144° = -\tau$$

where τ is the "golden ratio" of antiquity which satisfies the equations:

$$\tau^2 = \tau + 1$$
and $$\frac{1}{\tau} = \tau - 1$$

The actual value of τ is $\frac{1}{2}(\sqrt{5} + 1) = 1.6180339\ldots$

3.   Values of f(R) for Various Operations

The quantity f(R) is the contribution to the character of the Cartesian representation by *each* atom *un*shifted by an operation.

Operation	f(R)	Operation	f(R)
E	3	S_3	-2
σ	1	S_4	-1
i	-3	S_5	$\tau - 2$
C_2	-1	S_5^3	$-1 - \tau$
C_3	0	S_5^7	$-1 - \tau$
C_4	1	S_5^9	$\tau - 2$
C_5	τ	S_6	0
C_5^2	$1 - \tau$		
C_5^3	$1 - \tau$	C_n^k	$1 + 2\cos(2\pi k/n)$
C_6	2	S_n^k	$-1 + 2\cos(2\pi k/n)$

Character Tables for Chemically Important Symmetry Groups

1. The Nonaxial Groups

C_1	E
A	1

C_s	E	σ_h		
A'	1	1	x, y, R_z	x^2, y^2, z^2, xy
A''	1	-1	z, R_x, R_y	yz, xz

C_i	E	i		
A_g	1	1	R_x, R_y, R_z	x^2, y^2, z^2 xy, xz, yz
A_u	1	-1	x, y, z	

2. The C_n Groups

C_2	E	C_2		
A	1	1	z, R_z	x^2, y^2, z^2, xy
B	1	-1	x, y, R_x, R_y	yz, xz

C_3	E	C_3	$C_3{}^2$		$\varepsilon = \exp(2\pi i/3)$
A	1	1	1	z, R_z	$x^2 + y^2, z^2$
E	$\begin{Bmatrix} 1 & \varepsilon & \varepsilon^* \\ 1 & \varepsilon^* & \varepsilon \end{Bmatrix}$			$(x, y)(R_x, R_y)$	$(x^2 - y^2, xy)(yz, xz)$

The C_n Groups (continued)

C_4	E	C_4	C_2	$C_4{}^3$		
A	1	1	1	1	z, R_z	$x^2 + y^2, z^2$
B	1	-1	1	-1		$x^2 - y^2, xy$
E	$\begin{cases} 1 \\ 1 \end{cases}$	$\begin{matrix} i \\ -i \end{matrix}$	$\begin{matrix} -1 \\ -1 \end{matrix}$	$\begin{matrix} -i \\ i \end{matrix}$	$(x, y)(R_x, R_y)$	(yz, xz)

C_5	E	C_5	$C_5{}^2$	$C_5{}^3$	$C_5{}^4$		$\varepsilon = \exp(2\pi i/5)$
A	1	1	1	1	1	z, R_z	$x^2 + y^2, z^2$
E_1	$\begin{cases} 1 \\ 1 \end{cases}$	$\begin{matrix} \varepsilon \\ \varepsilon^* \end{matrix}$	$\begin{matrix} \varepsilon^2 \\ \varepsilon^{2*} \end{matrix}$	$\begin{matrix} \varepsilon^{2*} \\ \varepsilon^2 \end{matrix}$	$\begin{matrix} \varepsilon^* \\ \varepsilon \end{matrix}$	$(x, y)(R_x, R_y)$	(yz, xz)
E_2	$\begin{cases} 1 \\ 1 \end{cases}$	$\begin{matrix} \varepsilon^2 \\ \varepsilon^{2*} \end{matrix}$	$\begin{matrix} \varepsilon^* \\ \varepsilon \end{matrix}$	$\begin{matrix} \varepsilon \\ \varepsilon^* \end{matrix}$	$\begin{matrix} \varepsilon^{2*} \\ \varepsilon^2 \end{matrix}$		$(x^2 - y^2, xy)$

C_6	E	C_6	C_3	C_2	$C_3{}^2$	$C_6{}^5$		$\varepsilon = \exp(2\pi i/6)$
A	1	1	1	1	1	1	z, R_z	$x^2 + y^2, z^2$
B	1	-1	1	-1	1	-1		
E_1	$\begin{cases} 1 \\ 1 \end{cases}$	$\begin{matrix} \varepsilon \\ \varepsilon^* \end{matrix}$	$\begin{matrix} -\varepsilon^* \\ -\varepsilon \end{matrix}$	$\begin{matrix} -1 \\ -1 \end{matrix}$	$\begin{matrix} -\varepsilon \\ -\varepsilon^* \end{matrix}$	$\begin{matrix} \varepsilon^* \\ \varepsilon \end{matrix}$	$\begin{matrix} (x, y) \\ (R_x, R_y) \end{matrix}$	(xz, yz)
E_2	$\begin{cases} 1 \\ 1 \end{cases}$	$\begin{matrix} -\varepsilon^* \\ -\varepsilon \end{matrix}$	$\begin{matrix} -\varepsilon \\ -\varepsilon^* \end{matrix}$	$\begin{matrix} 1 \\ 1 \end{matrix}$	$\begin{matrix} -\varepsilon^* \\ -\varepsilon \end{matrix}$	$\begin{matrix} -\varepsilon \\ -\varepsilon^* \end{matrix}$		$(x^2 - y^2, xy)$

C_7	E	C_7	$C_7{}^2$	$C_7{}^3$	$C_7{}^4$	$C_7{}^5$	$C_7{}^6$		$\varepsilon = \exp(2\pi i/7)$
A	1	1	1	1	1	1	1	z, R_z	$x^2 + y^2, z^2$
E_1	$\begin{cases} 1 \\ 1 \end{cases}$	$\begin{matrix} \varepsilon \\ \varepsilon^* \end{matrix}$	$\begin{matrix} \varepsilon^2 \\ \varepsilon^{2*} \end{matrix}$	$\begin{matrix} \varepsilon^3 \\ \varepsilon^{3*} \end{matrix}$	$\begin{matrix} \varepsilon^{3*} \\ \varepsilon^3 \end{matrix}$	$\begin{matrix} \varepsilon^{2*} \\ \varepsilon^2 \end{matrix}$	$\begin{matrix} \varepsilon^* \\ \varepsilon \end{matrix}$	$\begin{matrix} (x, y) \\ (R_x, R_y) \end{matrix}$	(xz, yz)
E_2	$\begin{cases} 1 \\ 1 \end{cases}$	$\begin{matrix} \varepsilon^2 \\ \varepsilon^{2*} \end{matrix}$	$\begin{matrix} \varepsilon^{3*} \\ \varepsilon^3 \end{matrix}$	$\begin{matrix} \varepsilon^* \\ \varepsilon \end{matrix}$	$\begin{matrix} \varepsilon \\ \varepsilon^* \end{matrix}$	$\begin{matrix} \varepsilon^3 \\ \varepsilon^{3*} \end{matrix}$	$\begin{matrix} \varepsilon^{2*} \\ \varepsilon^2 \end{matrix}$		$(x^2 - y^2, xy)$
E_3	$\begin{cases} 1 \\ 1 \end{cases}$	$\begin{matrix} \varepsilon^3 \\ \varepsilon^{3*} \end{matrix}$	$\begin{matrix} \varepsilon^* \\ \varepsilon \end{matrix}$	$\begin{matrix} \varepsilon^2 \\ \varepsilon^{2*} \end{matrix}$	$\begin{matrix} \varepsilon^{2*} \\ \varepsilon^2 \end{matrix}$	$\begin{matrix} \varepsilon \\ \varepsilon^* \end{matrix}$	$\begin{matrix} \varepsilon^{3*} \\ \varepsilon^3 \end{matrix}$		

C_8	E	C_8	C_4	C_2	$C_4{}^3$	$C_8{}^3$	$C_8{}^5$	$C_8{}^7$		$\varepsilon = \exp(2\pi i/8)$
A	1	1	1	1	1	1	1	1	z, R_z	$x^2 + y^2, z^2$
B	1	-1	1	1	1	-1	-1	-1		
E_1	$\begin{cases} 1 \\ 1 \end{cases}$	$\begin{matrix} \varepsilon \\ \varepsilon^* \end{matrix}$	$\begin{matrix} i \\ -i \end{matrix}$	$\begin{matrix} -1 \\ -1 \end{matrix}$	$\begin{matrix} -i \\ i \end{matrix}$	$\begin{matrix} -\varepsilon^* \\ -\varepsilon \end{matrix}$	$\begin{matrix} -\varepsilon \\ -\varepsilon^* \end{matrix}$	$\begin{matrix} \varepsilon^* \\ \varepsilon \end{matrix}$	$\begin{matrix} (x, y) \\ (R_x, R_y) \end{matrix}$	(xz, yz)
E_2	$\begin{cases} 1 \\ 1 \end{cases}$	$\begin{matrix} i \\ -i \end{matrix}$	$\begin{matrix} -1 \\ -1 \end{matrix}$	$\begin{matrix} 1 \\ 1 \end{matrix}$	$\begin{matrix} -1 \\ -1 \end{matrix}$	$\begin{matrix} -i \\ i \end{matrix}$	$\begin{matrix} i \\ -i \end{matrix}$	$\begin{matrix} -i \\ i \end{matrix}$		$(x^2 - y^2, xy)$
E_3	$\begin{cases} 1 \\ 1 \end{cases}$	$\begin{matrix} -\varepsilon \\ -\varepsilon^* \end{matrix}$	$\begin{matrix} i \\ -i \end{matrix}$	$\begin{matrix} -1 \\ -1 \end{matrix}$	$\begin{matrix} -i \\ i \end{matrix}$	$\begin{matrix} \varepsilon^* \\ \varepsilon \end{matrix}$	$\begin{matrix} \varepsilon \\ \varepsilon^* \end{matrix}$	$\begin{matrix} -\varepsilon^* \\ -\varepsilon \end{matrix}$		

3. The D_n Groups

D_2	E	$C_2(z)$	$C_2(y)$	$C_2(x)$		
A	1	1	1	1		x^2, y^2, z^2
B_1	1	1	-1	-1	z, R_z	xy
B_2	1	-1	1	-1	y, R_y	xz
B_3	1	-1	-1	1	x, R_x	yz

D_3	E	$2C_3$	$3C_2$		
A_1	1	1	1		$x^2 + y^2, z^2$
A_2	1	1	-1	z, R_z	
E	2	-1	0	$(x, y)(R_x, R_y)$	$(x^2 - y^2, xy)(xz, yz)$

D_4	E	$2C_4$	$C_2(=C_4{}^2)$	$2C_2'$	$2C_2''$		
A_1	1	1	1	1	1		$x^2 + y^2, z^2$
A_2	1	1	1	-1	-1	z, R_z	
B_1	1	-1	1	1	-1		$x^2 - y^2$
B_2	1	-1	1	-1	1		xy
E	2	0	-2	0	0	$(x, y)(R_x, R_y)$	(xz, yz)

D_5	E	$2C_5$	$2C_5{}^2$	$5C_2$		
A_1	1	1	1	1		$x^2 + y^2, z^2$
A_2	1	1	1	-1	z, R_z	
E_1	2	$2\cos 72°$	$2\cos 144°$	0	$(x, y)(R_x, R_y)$	(xz, yz)
E_2	2	$2\cos 144°$	$2\cos 72°$	0		$(x^2 - y^2, xy)$

D_6	E	$2C_6$	$2C_3$	C_2	$3C_2'$	$3C_2''$		
A_1	1	1	1	1	1	1		$x^2 + y^2, z^2$
A_2	1	1	1	1	-1	-1	z, R_z	
B_1	1	-1	1	-1	1	-1		
B_2	1	-1	1	-1	-1	1		
E_1	2	1	-1	-2	0	0	$(x, y)(R_x, R_y)$	(xz, yz)
E_2	2	-1	-1	2	0	0		$(x^2 - y^2, xy)$

4. The C_{nv} Groups

C_{2v}	E	C_2	$\sigma_v(xz)$	$\sigma_v'(yz)$		
A_1	1	1	1	1	z	x^2, y^2, z^2
A_2	1	1	-1	-1	R_z	xy
B_1	1	-1	1	-1	x, R_y	xz
B_2	1	-1	-1	1	y, R_x	yz

C_{3v}	E	$2C_3$	$3\sigma_v$		
A_1	1	1	1	z	$x^2 + y^2, z^2$
A_2	1	1	-1	R_z	
E	2	-1	0	$(x, y)(R_x, R_y)$	$(x^2 - y^2, xy)(xz, yz)$

C_{4v}	E	$2C_4$	C_2	$2\sigma_v$	$2\sigma_d$		
A_1	1	1	1	1	1	z	$x^2 + y^2, z^2$
A_2	1	1	1	-1	-1	R_z	
B_1	1	-1	1	1	-1		$x^2 - y^2$
B_2	1	-1	1	-1	1		xy
E	2	0	-2	0	0	$(x, y)(R_x, R_y)$	(xz, yz)

C_{5v}	E	$2C_5$	$2C_5^2$	$5\sigma_v$		
A_1	1	1	1	1	z	$x^2 + y^2, z^2$
A_2	1	1	1	-1	R_z	
E_1	2	$2 \cos 72°$	$2 \cos 144°$	0	$(x, y)(R_x, R_y)$	(xz, yz)
E_2	2	$2 \cos 144°$	$2 \cos 72°$	0		$(x^2 - y^2, xy)$

C_{6v}	E	$2C_6$	$2C_3$	C_2	$3\sigma_v$	$3\sigma_d$		
A_1	1	1	1	1	1	1	z	$x^2 + y^2, z^2$
A_2	1	1	1	1	-1	-1	R_z	
B_1	1	-1	1	-1	1	-1		
B_2	1	-1	1	-1	-1	1		
E_1	2	1	-1	-2	0	0	$(x, y)(R_x, R_y)$	(xz, yz)
E_2	2	-1	-1	2	0	0		$(x^2 - y^2, xy)$

5. The C_{nh} Groups

C_{2h}	E	C_2	i	σ_h		
A_g	1	1	1	1	R_z	x^2, y^2, z^2, xy
B_g	1	-1	1	-1	R_x, R_y	xz, yz
A_u	1	1	-1	-1	z	
B_u	1	-1	-1	1	x, y	

C_{3h}	E	C_3	$C_3{}^2$	σ_h	S_3	$S_3{}^5$			$\varepsilon = \exp(2\pi i/3)$
A'	1	1	1	1	1	1	R_z	x^2+y^2, z^2	
E'	$\begin{Bmatrix}1\\1\end{Bmatrix}$	$\begin{matrix}\varepsilon\\\varepsilon^*\end{matrix}$	$\begin{matrix}\varepsilon^*\\\varepsilon\end{matrix}$	$\begin{matrix}1\\1\end{matrix}$	$\begin{matrix}\varepsilon\\\varepsilon^*\end{matrix}$	$\begin{matrix}\varepsilon^*\\\varepsilon\end{matrix}$	(x,y)	(x^2-y^2, xy)	
A''	1	1	1	-1	-1	-1	z		
E''	$\begin{Bmatrix}1\\1\end{Bmatrix}$	$\begin{matrix}\varepsilon\\\varepsilon^*\end{matrix}$	$\begin{matrix}\varepsilon^*\\\varepsilon\end{matrix}$	$\begin{matrix}-1\\-1\end{matrix}$	$\begin{matrix}-\varepsilon\\-\varepsilon^*\end{matrix}$	$\begin{matrix}-\varepsilon^*\\\varepsilon\end{matrix}$	(R_x, R_y)	(xz, yz)	

C_{4h}	E	C_4	C_2	$C_4{}^3$	i	$S_4{}^3$	σ_h	S_4		
A_g	1	1	1	1	1	1	1	1	R_z	x^2+y^2, z^2
B_g	1	-1	1	-1	1	-1	1	-1		x^2-y^2, xy
E_g	$\begin{Bmatrix}1\\1\end{Bmatrix}$	$\begin{matrix}i\\-i\end{matrix}$	$\begin{matrix}-1\\-1\end{matrix}$	$\begin{matrix}-i\\i\end{matrix}$	$\begin{matrix}1\\1\end{matrix}$	$\begin{matrix}i\\-i\end{matrix}$	$\begin{matrix}-1\\-1\end{matrix}$	$\begin{matrix}-i\\i\end{matrix}$	(R_x, R_y)	(xz, yz)
A_u	1	1	1	1	-1	-1	-1	-1	z	
B_u	1	-1	1	-1	-1	1	-1	1		
E_u	$\begin{Bmatrix}1\\1\end{Bmatrix}$	$\begin{matrix}i\\-i\end{matrix}$	$\begin{matrix}-1\\-1\end{matrix}$	$\begin{matrix}-i\\i\end{matrix}$	$\begin{matrix}-1\\-1\end{matrix}$	$\begin{matrix}-i\\i\end{matrix}$	$\begin{matrix}1\\1\end{matrix}$	$\begin{matrix}i\\-i\end{matrix}$	(x,y)	

C_{5h}	E	C_5	$C_5{}^2$	$C_5{}^3$	$C_5{}^4$	σ_h	S_5	$S_5{}^7$	$S_5{}^3$	$S_5{}^9$			$\varepsilon = \exp(2\pi i/5)$
A'	1	1	1	1	1	1	1	1	1	1	R_z	x^2+y^2, z^2	
E_1'	$\begin{Bmatrix}1\\1\end{Bmatrix}$	$\begin{matrix}\varepsilon\\\varepsilon^*\end{matrix}$	$\begin{matrix}\varepsilon^2\\\varepsilon^{2*}\end{matrix}$	$\begin{matrix}\varepsilon^{2*}\\\varepsilon^2\end{matrix}$	$\begin{matrix}\varepsilon^*\\\varepsilon\end{matrix}$	$\begin{matrix}1\\1\end{matrix}$	$\begin{matrix}\varepsilon\\\varepsilon^*\end{matrix}$	$\begin{matrix}\varepsilon^2\\\varepsilon^{2*}\end{matrix}$	$\begin{matrix}\varepsilon^{2*}\\\varepsilon^2\end{matrix}$	$\begin{matrix}\varepsilon^*\\\varepsilon\end{matrix}$	(x,y)		
E_2'	$\begin{Bmatrix}1\\1\end{Bmatrix}$	$\begin{matrix}\varepsilon^2\\\varepsilon^{2*}\end{matrix}$	$\begin{matrix}\varepsilon^*\\\varepsilon\end{matrix}$	$\begin{matrix}\varepsilon\\\varepsilon^*\end{matrix}$	$\begin{matrix}\varepsilon^{2*}\\\varepsilon^2\end{matrix}$	$\begin{matrix}1\\1\end{matrix}$	$\begin{matrix}\varepsilon^2\\\varepsilon^{2*}\end{matrix}$	$\begin{matrix}\varepsilon^*\\\varepsilon\end{matrix}$	$\begin{matrix}\varepsilon\\\varepsilon^*\end{matrix}$	$\begin{matrix}\varepsilon^{2*}\\\varepsilon^2\end{matrix}$		(x^2-y^2, xy)	
A''	1	1	1	1	1	-1	-1	-1	-1	-1	z		
E_1''	$\begin{Bmatrix}1\\1\end{Bmatrix}$	$\begin{matrix}\varepsilon\\\varepsilon^*\end{matrix}$	$\begin{matrix}\varepsilon^2\\\varepsilon^{2*}\end{matrix}$	$\begin{matrix}\varepsilon^{2*}\\\varepsilon^2\end{matrix}$	$\begin{matrix}\varepsilon^*\\\varepsilon\end{matrix}$	$\begin{matrix}-1\\-1\end{matrix}$	$\begin{matrix}-\varepsilon\\-\varepsilon^*\end{matrix}$	$\begin{matrix}-\varepsilon^2\\-\varepsilon^{2*}\end{matrix}$	$\begin{matrix}-\varepsilon^{2*}\\-\varepsilon^2\end{matrix}$	$\begin{matrix}-\varepsilon^*\\-\varepsilon\end{matrix}$	(R_x, R_y)	(xz, yz)	
E_2''	$\begin{Bmatrix}1\\1\end{Bmatrix}$	$\begin{matrix}\varepsilon^2\\\varepsilon^{2*}\end{matrix}$	$\begin{matrix}\varepsilon^*\\\varepsilon\end{matrix}$	$\begin{matrix}\varepsilon\\\varepsilon^*\end{matrix}$	$\begin{matrix}\varepsilon^{2*}\\\varepsilon^2\end{matrix}$	$\begin{matrix}-1\\-1\end{matrix}$	$\begin{matrix}-\varepsilon^2\\-\varepsilon^{2*}\end{matrix}$	$\begin{matrix}-\varepsilon^*\\-\varepsilon\end{matrix}$	$\begin{matrix}-\varepsilon\\-\varepsilon^*\end{matrix}$	$\begin{matrix}-\varepsilon^{2*}\\-\varepsilon^2\end{matrix}$			

C_{6h}	E	C_6	C_3	C_2	$C_3{}^2$	$C_6{}^5$	i	$S_3{}^5$	$S_6{}^5$	σ_h	S_6	S_3			$\varepsilon = \exp(2\pi i/6)$
A_g	1	1	1	1	1	1	1	1	1	1	1	1	R_z	x^2+y^2, z^2	
B_g	1	-1	1	-1	1	-1	1	-1	1	-1	1	-1			
E_{1g}	$\begin{Bmatrix}1\\1\end{Bmatrix}$	$\begin{matrix}\varepsilon\\\varepsilon^*\end{matrix}$	$\begin{matrix}-\varepsilon^*\\-\varepsilon\end{matrix}$	$\begin{matrix}-1\\-1\end{matrix}$	$\begin{matrix}-\varepsilon\\-\varepsilon^*\end{matrix}$	$\begin{matrix}\varepsilon^*\\\varepsilon\end{matrix}$	$\begin{matrix}1\\1\end{matrix}$	$\begin{matrix}\varepsilon\\\varepsilon^*\end{matrix}$	$\begin{matrix}-\varepsilon^*\\-\varepsilon\end{matrix}$	$\begin{matrix}-1\\-1\end{matrix}$	$\begin{matrix}-\varepsilon\\-\varepsilon^*\end{matrix}$	$\begin{matrix}\varepsilon^*\\\varepsilon\end{matrix}$	(R_x, R_y)	(xz, yz)	
E_{2g}	$\begin{Bmatrix}1\\1\end{Bmatrix}$	$\begin{matrix}-\varepsilon^*\\-\varepsilon\end{matrix}$	$\begin{matrix}-\varepsilon\\-\varepsilon^*\end{matrix}$	$\begin{matrix}1\\1\end{matrix}$	$\begin{matrix}-\varepsilon^*\\-\varepsilon\end{matrix}$	$\begin{matrix}-\varepsilon\\-\varepsilon^*\end{matrix}$	$\begin{matrix}1\\1\end{matrix}$	$\begin{matrix}-\varepsilon^*\\-\varepsilon\end{matrix}$	$\begin{matrix}-\varepsilon\\-\varepsilon^*\end{matrix}$	$\begin{matrix}1\\1\end{matrix}$	$\begin{matrix}-\varepsilon^*\\-\varepsilon\end{matrix}$	$\begin{matrix}-\varepsilon\\-\varepsilon^*\end{matrix}$		(x^2-y^2, xy)	
A_u	1	1	1	1	1	1	-1	-1	-1	-1	-1	-1	z		
B_u	1	-1	1	-1	1	-1	-1	1	-1	1	-1	1			
E_{1u}	$\begin{Bmatrix}1\\1\end{Bmatrix}$	$\begin{matrix}\varepsilon\\\varepsilon^*\end{matrix}$	$\begin{matrix}-\varepsilon^*\\-\varepsilon\end{matrix}$	$\begin{matrix}-1\\-1\end{matrix}$	$\begin{matrix}-\varepsilon\\-\varepsilon^*\end{matrix}$	$\begin{matrix}\varepsilon^*\\\varepsilon\end{matrix}$	$\begin{matrix}-1\\-1\end{matrix}$	$\begin{matrix}-\varepsilon\\-\varepsilon^*\end{matrix}$	$\begin{matrix}\varepsilon^*\\\varepsilon\end{matrix}$	$\begin{matrix}1\\1\end{matrix}$	$\begin{matrix}\varepsilon\\\varepsilon^*\end{matrix}$	$\begin{matrix}-\varepsilon^*\\-\varepsilon\end{matrix}$	(x,y)		
E_{2u}	$\begin{Bmatrix}1\\1\end{Bmatrix}$	$\begin{matrix}-\varepsilon^*\\-\varepsilon\end{matrix}$	$\begin{matrix}-\varepsilon\\-\varepsilon^*\end{matrix}$	$\begin{matrix}1\\1\end{matrix}$	$\begin{matrix}-\varepsilon^*\\-\varepsilon\end{matrix}$	$\begin{matrix}-\varepsilon\\-\varepsilon^*\end{matrix}$	$\begin{matrix}-1\\-1\end{matrix}$	$\begin{matrix}\varepsilon^*\\\varepsilon\end{matrix}$	$\begin{matrix}\varepsilon\\\varepsilon^*\end{matrix}$	$\begin{matrix}-1\\-1\end{matrix}$	$\begin{matrix}\varepsilon^*\\\varepsilon\end{matrix}$	$\begin{matrix}\varepsilon\\\varepsilon^*\end{matrix}$			

6. The D_{nh} Groups

D_{2h}	E	$C_2(z)$	$C_2(y)$	$C_2(x)$	i	$\sigma(xy)$	$\sigma(xz)$	$\sigma(yz)$		
A_g	1	1	1	1	1	1	1	1		x^2, y^2, z^2
B_{1g}	1	1	-1	-1	1	1	-1	-1	R_z	xy
B_{2g}	1	-1	1	-1	1	-1	1	-1	R_y	xz
B_{3g}	1	-1	-1	1	1	-1	-1	1	R_x	yz
A_u	1	1	1	1	-1	-1	-1	-1		
B_{1u}	1	1	-1	-1	-1	-1	1	1	z	
B_{2u}	1	-1	1	-1	-1	1	-1	1	y	
B_{3u}	1	-1	-1	1	-1	1	1	-1	x	

D_{3h}	E	$2C_3$	$3C_2$	σ_h	$2S_3$	$3\sigma_v$		
A_1'	1	1	1	1	1	1		$x^2 + y^2, z^2$
A_2'	1	1	-1	1	1	-1	R_z	
E'	2	-1	0	2	-1	0	(x, y)	$(x^2 - y^2, xy)$
A_1''	1	1	1	-1	-1	1		
A_2''	1	1	-1	-1	-1	1	z	
E''	2	-1	0	-2	1	0	(R_x, R_y)	(xz, yz)

D_{4h}	E	$2C_4$	C_2	$2C_2'$	$2C_2''$	i	$2S_4$	σ_h	$2\sigma_r$	$2\sigma_d$		
A_{1g}	1	1	1	1	1	1	1	1	1	1		$x^2 + y^2, z^2$
A_{2g}	1	1	1	-1	-1	1	1	1	-1	-1	R_z	
B_{1g}	1	-1	1	1	-1	1	-1	1	1	-1		$x^2 - y^2$
B_{2g}	1	-1	1	-1	1	1	-1	1	-1	1		xy
E_g	2	0	-2	0	0	2	0	-2	0	0	(R_x, R_y)	(xz, yz)
A_{1u}	1	1	1	1	1	-1	-1	-1	-1	-1		
A_{2u}	1	1	1	-1	-1	-1	-1	-1	1	1	z	
B_{1u}	1	-1	1	1	-1	-1	1	-1	-1	1		
B_{2u}	1	-1	1	-1	1	-1	1	-1	1	-1		
E_u	2	0	-2	0	0	-2	0	2	0	0	(x, y)	

D_{5h}	E	$2C_5$	$2C_5^2$	$5C_2$	σ_h	$2S_5$	$2S_5^3$	$5\sigma_v$		
A_1'	1	1	1	1	1	1	1	1		$x^2 + y^2, z^2$
A_2'	1	1	1	-1	1	1	1	-1	R_z	
E_1'	2	$2\cos 72°$	$2\cos 144°$	0	2	$2\cos 72°$	$2\cos 144°$	0	(x, y)	
E_2'	2	$2\cos 144°$	$2\cos 72°$	0	2	$2\cos 144°$	$2\cos 72°$	0		$(x^2 - y^2, xy)$
A_1''	1	1	1	1	-1	-1	-1	-1		
A_2''	1	1	1	-1	-1	-1	-1	1	z	
E_1''	2	$2\cos 72°$	$2\cos 144°$	0	-2	$-2\cos 72°$	$-2\cos 144°$	0	(R_x, R_y)	(xz, yz)
E_2''	2	$2\cos 144°$	$2\cos 72°$	0	-2	$-2\cos 144°$	$-2\cos 72°$	0		

D_{6h}	E	$2C_6$	$2C_3$	C_2	$3C_2'$	$3C_2''$	i	$2S_3$	$2S_6$	σ_h	$3\sigma_d$	$3\sigma_v$		
A_{1g}	1	1	1	1	1	1	1	1	1	1	1	1		$x^2 + y^2, z^2$
A_{2g}	1	1	1	1	-1	-1	1	1	1	1	-1	-1	R_z	
B_{1g}	1	-1	1	-1	1	-1	1	-1	1	-1	1	-1		
B_{2g}	1	-1	1	-1	-1	1	1	-1	1	-1	-1	1		
E_{1g}	2	1	-1	-2	0	0	2	1	-1	-2	0	0	(R_x, R_y)	(xz, yz)
E_{2g}	2	-1	-1	2	0	0	2	-1	-1	2	0	0		$(x^2 - y^2, xy)$
A_{1u}	1	1	1	1	1	1	-1	-1	-1	-1	-1	-1		
A_{2u}	1	1	1	1	-1	-1	-1	-1	-1	-1	1	1	z	
B_{1u}	1	-1	1	-1	1	-1	-1	1	-1	1	-1	1		
B_{2u}	1	-1	1	-1	-1	1	-1	1	-1	1	1	-1		
E_{1u}	2	1	-1	-2	0	0	-2	-1	1	2	0	0	(x, y)	
E_{2u}	2	-1	-1	2	0	0	-2	1	1	-2	0	0		

D_{8h}	E	$2C_8$	$2C_8^3$	$2C_4$	C_2	$4C_2'$	$4C_2''$	i	$2S_8$	$2S_8^3$	$2S_4$	σ_h	$4\sigma_d$	$4\sigma_v$		
A_{1g}	1	1	1	1	1	1	1	1	1	1	1	1	1	1		x^2+y^2, z^2
A_{2g}	1	1	1	1	1	-1	-1	1	1	1	1	1	-1	-1	R_z	
B_{1g}	1	-1	-1	1	1	1	-1	1	-1	-1	1	1	1	-1		
B_{2g}	1	-1	-1	1	1	-1	1	1	-1	-1	1	1	-1	1		
E_{1g}	2	$\sqrt{2}$	$-\sqrt{2}$	0	-2	0	0	2	$\sqrt{2}$	$-\sqrt{2}$	0	-2	0	0	(R_x, R_y)	(xz, yz)
E_{2g}	2	0	0	-2	2	0	0	2	0	0	-2	2	0	0		(x^2-y^2, xy)
E_{3g}	2	$-\sqrt{2}$	$\sqrt{2}$	0	-2	0	0	2	$-\sqrt{2}$	$\sqrt{2}$	0	-2	0	0		
A_{1u}	1	1	1	1	1	1	1	-1	-1	-1	-1	-1	-1	-1		
A_{2u}	1	1	1	1	1	-1	-1	-1	-1	-1	-1	-1	1	1	z	
B_{1u}	1	-1	-1	1	1	1	-1	-1	1	1	-1	-1	-1	1		
B_{2u}	1	-1	-1	1	1	-1	1	-1	1	1	-1	-1	1	-1		
E_{1u}	2	$\sqrt{2}$	$-\sqrt{2}$	0	-2	0	0	-2	$-\sqrt{2}$	$\sqrt{2}$	0	2	0	0	(x, y)	
E_{2u}	2	0	0	-2	2	0	0	-2	0	0	2	-2	0	0		
E_{3u}	2	$-\sqrt{2}$	$\sqrt{2}$	0	-2	0	0	-2	$\sqrt{2}$	$-\sqrt{2}$	0	2	0	0		

7. The D_{nd} Groups

D_{2d}	E	$2S_4$	C_2	$2C_2'$	$2\sigma_d$		
A_1	1	1	1	1	1		x^2+y^2, z^2
A_2	1	1	1	-1	-1	R_z	
B_1	1	-1	1	1	-1		x^2-y^2
B_2	1	-1	1	-1	1	z	xy
E	2	0	-2	0	0	(x,y); (R_x, R_y)	(xz, yz)

D_{3d}	E	$2C_3$	$3C_2$	i	$2S_6$	$3\sigma_d$		
A_{1g}	1	1	1	1	1	1		x^2+y^2, z^2
A_{2g}	1	1	-1	1	1	-1	R_z	
E_g	2	-1	0	2	-1	0	(R_x, R_y)	$(x^2-y^2, xy),$ (xz, yz)
A_{1u}	1	1	1	-1	-1	-1		
A_{2u}	1	1	-1	-1	-1	1	z	
E_u	2	-1	0	-2	1	0	(x,y)	

D_{4d}	E	$2S_8$	$2C_4$	$2S_8^3$	C_2	$4C_2'$	$4\sigma_d$		
A_1	1	1	1	1	1	1	1		x^2+y^2, z^2
A_2	1	1	1	1	1	-1	-1	R_z	
B_1	1	-1	1	-1	1	1	-1		
B_2	1	-1	1	-1	1	-1	1	z	
E_1	2	$\sqrt{2}$	0	$-\sqrt{2}$	-2	0	0	(x, y)	
E_2	2	0	-2	0	2	0	0		(x^2-y^2, xy)
E_3	2	$-\sqrt{2}$	0	$\sqrt{2}$	-2	0	0	(R_x, R_y)	(xz, yz)

D_{5d}	E	$2C_5$	$2C_5^2$	$5C_2$	i	$2S_{10}^3$	$2S_{10}$	$5\sigma_d$		
A_{1g}	1	1	1	1	1	1	1	1		x^2+y^2, z^2
A_{2g}	1	1	1	-1	1	1	1	-1	R_z	
E_{1g}	2	$2\cos 72°$	$2\cos 144°$	0	2	$2\cos 72°$	$2\cos 144°$	0	(R_x, R_y)	(xz, yz)
E_{2g}	2	$2\cos 144°$	$2\cos 72°$	0	2	$2\cos 144°$	$2\cos 72°$	0		(x^2-y^2, xy)
A_{1u}	1	1	1	1	-1	-1	-1	-1		
A_{2u}	1	1	1	-1	-1	-1	-1	1	z	
E_{1u}	2	$2\cos 72°$	$2\cos 144°$	0	-2	$-2\cos 72°$	$-2\cos 144°$	0	(x, y)	
E_{2u}	2	$2\cos 144°$	$2\cos 72°$	0	-2	$-2\cos 144°$	$-2\cos 72°$	0		

The D_{nd} Groups (*continued*)

D_{6d}	E	$2S_{12}$	$2C_6$	$2S_4$	$2C_3$	$2S_{12}^5$	C_2	$6C_2'$	$6\sigma_d$		
A_1	1	1	1	1	1	1	1	1	1		x^2+y^2, z^2
A_2	1	1	1	1	1	1	1	-1	-1	R_z	
B_1	1	-1	1	-1	1	1	1	1	-1		
B_2	1	-1	1	-1	1	-1	1	-1	1	z	
E_1	2	$\sqrt{3}$	1	0	-1	$-\sqrt{3}$	-2	0	0	(x,y)	
E_2	2	1	-1	-2	-1	1	2	0	0		(x^2-y^2, xy)
E_3	2	0	-2	0	2	0	-2	0	0		
E_4	2	-1	-1	2	-1	-1	2	0	0		
E_5	2	$-\sqrt{3}$	1	0	-1	$\sqrt{3}$	-2	0	0	(R_x, R_y)	(xz, yz)

8. The S_n Groups

S_4	E	S_4	C_2	S_4^3		
A	1	1	1	1	R_z	x^2+y^2, z^2
B	1	-1	1	-1	z	x^2-y^2, xy
E	$\begin{cases} 1 & i & -1 & -i \\ 1 & -i & -1 & i \end{cases}$				$(x,y); (R_x, R_y)$	(xz, yz)

S_6	E	C_3	C_3^2	i	S_6^5	S_6		$\varepsilon = \exp(2\pi i/3)$
A_g	1	1	1	1	1	1	R_z	x^2+y^2, z^2
E_g	$\begin{cases} 1 & \varepsilon & \varepsilon^* & 1 & \varepsilon & \varepsilon^* \\ 1 & \varepsilon^* & \varepsilon & 1 & \varepsilon^* & \varepsilon \end{cases}$						(R_x, R_y)	$(x^2-y^2, xy);$ (xz, yz)
A_u	1	1	1	-1	-1	-1	z	
E_u	$\begin{cases} 1 & \varepsilon & \varepsilon^* & -1 & -\varepsilon & -\varepsilon^* \\ 1 & \varepsilon^* & \varepsilon & -1 & -\varepsilon^* & -\varepsilon \end{cases}$						(x,y)	

S_8	E	S_8	C_4	S_8^3	C_2	S_8^5	C_4^3	S_8^7		$\varepsilon = \exp(2\pi i/8)$
A	1	1	1	1	1	1	1	1	R_z	x^2+y^2, z^2
B	1	-1	1	-1	1	-1	1	-1	z	
E_g	$\begin{cases} 1 & \varepsilon & i & -\varepsilon^* & -1 & -\varepsilon & -i & \varepsilon^* \\ 1 & \varepsilon^* & -i & -\varepsilon & -1 & -\varepsilon^* & i & \varepsilon \end{cases}$								$(x,y);$ (R_x, R_y)	
E_g	$\begin{cases} 1 & i & -1 & -i & 1 & i & -1 & -i \\ 1 & -i & -1 & i & 1 & -i & -1 & i \end{cases}$									(x^2-y^2, xy)
E_g	$\begin{cases} 1 & -\varepsilon^* & -i & \varepsilon & -1 & \varepsilon^* & i & -\varepsilon \\ 1 & -\varepsilon & i & \varepsilon^* & -1 & \varepsilon & -i & -\varepsilon^* \end{cases}$									(xz, yz)

9. The Cubic Groups

T	E	$4C_3$	$4C_3^2$	$3C_2$		$\varepsilon = \exp(2\pi i/3)$
A	1	1	1	1		$x^2+y^2+z^2$
E	$\begin{cases} 1 & \varepsilon & \varepsilon^* & 1 \\ 1 & \varepsilon^* & \varepsilon & 1 \end{cases}$					$(2z^2-x^2-y^2,$ $x^2-y^2)$
T	3	0	0	-1	$(R_x, R_y, R_z); (x,y,z)$	(xy, xz, yz)

The Cubic Groups (*continued*)

T_h	E	$4C_3$	$4C_3{}^2$	$3C_2$	i	$4S_6$	$4S_6{}^5$	$3\sigma_h$		$\varepsilon = \exp(2\pi i/3)$
A_g	1	1	1	1	1	1	1	1		$x^2+y^2+z^2$
A_u	1	1	1	1	-1	-1	-1	-1		
E_g	$\begin{cases}1\\1\end{cases}$	$\begin{matrix}\varepsilon\\\varepsilon^*\end{matrix}$	$\begin{matrix}\varepsilon^*\\\varepsilon\end{matrix}$	$\begin{matrix}1\\1\end{matrix}$	$\begin{matrix}1\\1\end{matrix}$	$\begin{matrix}\varepsilon\\\varepsilon^*\end{matrix}$	$\begin{matrix}\varepsilon^*\\\varepsilon\end{matrix}$	$\begin{matrix}1\\1\end{matrix}$		$(2z^2-x^2-y^2,\\x^2-y^2)$
E_u	$\begin{cases}1\\1\end{cases}$	$\begin{matrix}\varepsilon\\\varepsilon^*\end{matrix}$	$\begin{matrix}\varepsilon^*\\\varepsilon\end{matrix}$	$\begin{matrix}1\\1\end{matrix}$	$\begin{matrix}-1\\-1\end{matrix}$	$\begin{matrix}-\varepsilon\\-\varepsilon^*\end{matrix}$	$\begin{matrix}-\varepsilon^*\\-\varepsilon\end{matrix}$	$\begin{matrix}-1\\-1\end{matrix}$		
T_g	3	0	0	-1	1	0	0	-1	(R_x, R_y, R_z)	(xz, yz, xy)
T_u	3	0	0	-1	-1	0	0	1	(x, y, z)	

T_d	E	$8C_3$	$3C_2$	$6S_4$	$6\sigma_d$		
A_1	1	1	1	1	1		$x^2+y^2+z^2$
A_2	1	1	1	-1	-1		
E	2	-1	2	0	0		$(2z^2-x^2-y^2,\\x^2-y^2)$
T_1	3	0	-1	1	-1	(R_x, R_y, R_z)	
T_2	3	0	-1	-1	1	(x, y, z)	(xy, xz, yz)

O	E	$6C_4$	$3C_2(=C_4{}^2)$	$8C_3$	$6C_2$		
A_1	1	1	1	1	1		$x^2+y^2+z^2$
A_2	1	-1	1	1	-1		
E	2	0	2	-1	0		$(2z^2-x^2-y^2,\\x^2-y^2)$
T_1	3	1	-1	0	-1	$(R_x, R_y, R_z); (x, y, z)$	
T_2	3	-1	-1	0	1		(xy, xz, yz)

O_h	E	$8C_3$	$6C_2$	$6C_4$	$3C_2(=C_4{}^2)$	i	$6S_4$	$8S_6$	$3\sigma_h$	$6\sigma_d$		
A_{1g}	1	1	1	1	1	1	1	1	1	1		$x^2+y^2+z^2$
A_{2g}	1	1	-1	-1	1	1	-1	1	1	-1		
E_g	2	-1	0	0	2	2	0	-1	2	0		$(2z^2-x^2-y\\x^2-y^2)$
T_{1g}	3	0	-1	1	-1	3	1	0	-1	-1	(R_x, R_y, R_z)	
T_{2g}	3	0	1	-1	-1	3	-1	0	-1	1		(xz, yz, xy)
A_{1u}	1	1	1	1	1	-1	-1	-1	-1	-1		
A_{2u}	1	1	-1	-1	1	-1	1	-1	-1	1		
E_u	2	-1	0	0	2	-2	0	1	-2	0		
T_{1u}	3	0	-1	1	-1	-3	-1	0	1	1	(x, y, z)	
T_{2u}	3	0	1	-1	-1	-3	1	0	1	-1		

10. The Groups $C_{\infty v}$ and $D_{\infty h}$ for Linear Molecules

$C_{\infty v}$	E	$2C_\infty{}^\Phi$	$\cdots$	$\infty\sigma_v$		
$A_1 \equiv \Sigma^+$	1	1	$\cdots$	1	z	x^2+y^2, z^2
$A_2 \equiv \Sigma^-$	1	1	$\cdots$	-1	R_z	
$E_1 \equiv \Pi$	2	$2\cos\Phi$	$\cdots$	0	$(x, y); (R_x, R_y)$	(xz, yz)
$E_2 \equiv \Delta$	2	$2\cos 2\Phi$	$\cdots$	0		(x^2-y^2, xy)
$E_3 \equiv \Phi$	2	$2\cos 3\Phi$	$\cdots$	0		
$\cdots$	$\cdots$	$\cdots$	$\cdots$			

$D_{\infty h}$	E	$2C_\infty{}^\Phi$	$\cdots$	$\infty\sigma_v$	i	$2S_\infty{}^\Phi$	$\cdots$	∞C_2		
$\Sigma_g{}^+$	1	1	$\cdots$	1	1	1	$\cdots$	1		x^2+y^2, z^2
$\Sigma_g{}^-$	1	1	$\cdots$	-1	1	1	$\cdots$	-1	R_z	
Π_g	2	$2\cos\Phi$	$\cdots$	0	2	$-2\cos\Phi$	$\cdots$	0	(R_x, R_y)	(xz, yz)
Δ_g	2	$2\cos 2\Phi$	$\cdots$	0	2	$2\cos 2\Phi$	$\cdots$	0		(x^2-y^2, xy)
$\cdots$	$\cdots$	$\cdots$	$\cdots$		$\cdots$	$\cdots$	$\cdots$	$\cdots$		
$\Sigma_u{}^+$	1	1	$\cdots$	1	-1	-1	$\cdots$	-1	z	
$\Sigma_u{}^-$	1	1	$\cdots$	-1	-1	-1	$\cdots$	1		
Π_u	2	$2\cos\Phi$	$\cdots$	0	-2	$2\cos\Phi$	$\cdots$	0	(x, y)	
Δ_u	2	$2\cos 2\Phi$	$\cdots$	0	-2	$-2\cos 2\Phi$	$\cdots$	0		
$\cdots$	$\cdots$	$\cdots$	$\cdots$		$\cdots$	$\cdots$	$\cdots$	$\cdots$		

11. The Icosahedral Groups*

I_h	E	$12C_5$	$12C_5{}^2$	$20C_3$	$15C_2$	i	$12S_{10}$	$12S_{10}{}^3$	$20S_6$	15σ		
A_g	1	1	1	1	1	1	1	1	1	1		$x^2+y^2+z^2$
T_{1g}	3	$\frac{1}{2}(1+\sqrt{5})$	$\frac{1}{2}(1-\sqrt{5})$	0	-1	3	$\frac{1}{2}(1-\sqrt{5})$	$\frac{1}{2}(1+\sqrt{5})$	0	-1	(R_x, R_y, R_z)	
T_{2g}	3	$\frac{1}{2}(1-\sqrt{5})$	$\frac{1}{2}(1+\sqrt{5})$	0	-1	3	$\frac{1}{2}(1+\sqrt{5})$	$\frac{1}{2}(1-\sqrt{5})$	0	-1		
G_g	4	-1	-1	1	0	4	-1	-1	1	0		
H_g	5	0	0	-1	1	5	0	0	-1	1		$(2z^2-x^2-y^2,$ $x^2-y^2,$ $xy, yz, zx)$
A_u	1	1	1	1	1	-1	-1	-1	-1	-1		
T_{1u}	3	$\frac{1}{2}(1+\sqrt{5})$	$\frac{1}{2}(1-\sqrt{5})$	0	-1	-3	$-\frac{1}{2}(1-\sqrt{5})$	$-\frac{1}{2}(1+\sqrt{5})$	0	1	(x, y, z)	
T_{2u}	3	$\frac{1}{2}(1-\sqrt{5})$	$\frac{1}{2}(1+\sqrt{5})$	0	-1	-3	$-\frac{1}{2}(1+\sqrt{5})$	$-\frac{1}{2}(1-\sqrt{5})$	0	1		
G_u	4	-1	-1	1	0	-4	1	1	-1	0		
H_u	5	0	0	-1	1	-5	0	0	1	-1		

* For the pure rotation group I, the outlined section in the upper left is the character table; the g subscripts should, of course, be dropped and (x, y, z) assigned to the T_1 representation.

Index